A PHILOSOPHER LOOKS AT HUMAN BEINGS

Why do we think ourselves superior to all other animals? Are we right to think so? In this book, Michael Ruse explores these questions in religion, science, and philosophy. Some people think that the world is an organism – and that humans, as its highest part, have a natural value (this view appeals particularly to people of religion). Others think that the world is a machine – and that we therefore have responsibility for making our own value judgments (including judgments about ourselves). Ruse provides a compelling analysis of these two rival views and the age-old conflict between them. In a wide-ranging and fascinating discussion, he draws on Darwinism and existentialism to argue that only the view that the world is a machine does justice to our humanity. This new series offers short and personal perspectives by expert thinkers on topics that we all encounter in our everyday lives.

MICHAEL RUSE is the former Lucyle T. Werkmeister Professor of Philosophy at Florida State University. Over his fifty-year career he has authored and co-edited over sixty books on topics ranging from the history and philosophy of science, especially evolutionary biology, to the philosophy of religion. They include *Can a Darwinian Be a Christian?* (Cambridge, 2001), *The Gaia Hypothesis: Science on a Pagan Planet* (2013), and *A Meaning to Life* (2019).

A Philosopher Looks at

In this series, philosophers offer a personal and philosophical exploration of a topic of general interest.

Books in the series

A PHILOSOPHER LOOKS AT

HUMAN
BEINGS

MICHAEL RUSE

CAMBRIDGE
UNIVERSITY PRESS

CAMBRIDGE
UNIVERSITY PRESS

University Printing House, Cambridge CB2 8BS, United Kingdom

One Liberty Plaza, 20th Floor, New York, NY 10006, USA

477 Williamstown Road, Port Melbourne, VIC 3207, Australia

314–321, 3rd Floor, Plot 3, Splendor Forum, Jasola District Centre, New Delhi – 110025, India

79 Anson Road, #06–04/06, Singapore 079906

Cambridge University Press is part of the University of Cambridge.

It furthers the University's mission by disseminating knowledge in the pursuit of education, learning, and research at the highest international levels of excellence.

www.cambridge.org
Information on this title: www.cambridge.org/9781108820431
DOI: 10.1017/9781108907057

First published 2021

Printed in the United Kingdom by TJ Books Limited, Padstow Cornwall

A catalogue record for this publication is available from the British Library.

ISBN 978-1-108-82043-1 Paperback

For my son Oliver
and all the other first responders during the
coronavirus pandemic

CONTENTS

CONTENTS

ACKNOWLEDGMENTS

Over the years, many people have helped me to develop the ideas expressed in this essay. Edward O. Wilson, at Harvard University, has been both friend and support, and his enthusiasm for ideas has long been an inspiration. Robert J. Richards, at the University of Chicago, and I have long shared an interest in the development of nineteenth-century evolutionary thought. Joe Cain, at University College London, and I have long shared an interest in the development of twentieth-century evolutionary thought. I often disagree with all three of these men. They have taught me what every philosopher comes to realize. You learn most from those who challenge your ideas. Mirroring the wonderful help and encouragement I got from older scholars when I was young, one of the greatest joys of my life has been passing on the torch from my generation to the next. I have been co-editing the Cambridge *Elements* series in the Philosophy of Biology with the much-younger Grant Ramsey, at Leuven University in Belgium. I am hugely grateful for his comments on this essay.

My editor at Cambridge University Press, Hilary Gaskin, is far more than just a professional acquaintance. I am much in her debt for all that she has done for me and for asking me to write this essay. I thank also my content manager Thomas Haynes and my copy-editor Steven Holt. As an author, I am always so grateful to the people who help turn my manuscript into a published book. Closer to home, I thank William and Lucyle Werkmeister for their endowment of my professorship at Florida State University and for the research funds that go with it. My love for my wife Lizzie, and my gratitude for her never-failing support, cannot be put into words.

The information in Chapter 5 about Sewall Wright's ordering of biological concepts comes from the Simpson Papers, held by the American Philosophical Society in Philadelphia. They are used with permission. I use the New International Version of the Bible.

Introduction

I am a human being. What do I, as a philosopher, have to say about this? If I were a physiologist, I would be interested in what makes us tick – how the various parts of the body interact and work together. If I were a sociologist, I would be interested in humans in groups – why are there churches and priests and imams and that sort of thing? I am a philosopher, so what am I interested in and why do I have special authority or knowledge to speak about such things? We'll pick up on the second part of the question as we go along – the proof of the pudding is in the eating – but I'll tell you what I am interested in. Why do we humans think we are so special? Do we have good reason for this, or is it just self-deception based on ignorance and arrogance or (perhaps) a fear that we are not so very special?

Let me speak for myself. I once spent a week in Zimbabwe, and I must confess that I came away with a liking for warthogs. They are certainly not the world's most beautiful animals, but perhaps that was part of the attraction. Out on safari – cameras not guns! – you would see them trotting along in their families, quite content. As the father of five kids, that appealed to me right there. And they weren't afraid. They would come right up to the Victoria Falls Hotel – the one that King George and Queen Elizabeth stayed in during their trip to Southern Africa in 1947 – and graze contentedly in the grounds. But my liking is accompanied with a kind of smugness. Truly, we humans are not as other animals, or rather there are no other animals comparable with us humans and we are at the top. Warthogs are just fine, but they should know their place. They do not belong at the high table – in a comfortable armchair, looking out across at the Falls, sipping a gin and tonic, boring my wife to tears as I go on about the Royal Family. "If you disapprove of them so much, why do you keep talking about them?"

Am I justified in thinking this about my high status? Try to discount the fact that, like all philosophers, I think I am superior to all others. The brightest people on campus. Return to reality. I love my cairn terriers – Scruffy McGruff and Duncan Donut – more than a lot of humans. But truly, other than my late headmaster, of whom I am prepared to think any ill – a sentiment fully reciprocated – I don't think my dogs are the equals of other humans. That's not just the prejudice of an Englishman against the Scots. It's a question of status, of value, of worth. So much so, that when I am confronted with an appallingly awful human being – like Jeffrey Dahmer, the murderer, who not only preferred to have (gay) sex with his victims after (rather than before) they were dead but who used to cut them up and eat them – my first inclination is to say that such people are not real human beings. They are inhuman.

This is the problem I am going to wrestle with in this short book. What is it that makes human beings special, if indeed they are? My first chapter shows what is to me an amazing paradox, that people who agree about absolutely nothing else – who are in fact often in violent and public disagreement – agree entirely on the high status of human beings. That is a given. This leads into the second chapter, preparing the way for trying to speak to the beliefs of those in the first chapter. Science is going to be the key, but before we get to the actual science, we must dig into the underlying metaphysical presumptions that people bring to their science. These, we shall see, are a crucial element in the stew. Third, we turn to the science, what one might call the dominant position or paradigm of today, Darwinian evolutionary theory. We shall find out what it has to say about humans. Fourth and fifth, we shall evaluate its conclusions by circling back to the metaphysical assumptions unveiled in Chapter 2, starting to approach understanding of the positions taken in Chapter 1. Sixth, I shall take up the question of progress, trying to see if the different approaches really are so very different.

This leads to the all-important seventh and eighth chapters, where I show that what has gone before is very much more than a sophisticated exercise in navel gazing. Here the personal and professional come together and I show why I very much wanted to write this book. I was born in the British Midlands in 1940, shortly after the beginning of the Second World War. My parents were Quakers, and it was within that faith I was raised. Quakers are pacifists. They also have no dogmas. This does not mean they have no strong beliefs. They can equal the Jesuits on this. Try them on the "inner light" or "that of God in every person." But you – and that means starting as children – with the helpful guidance of older Friends (Quakers), must work these things out for yourself. Unlike the First World

War, the Second World War was a "good" war, in that it had to be fought. Hitler had to be stopped. Can one be a pacifist in such a situation? Should one be a pacifist in such a situation? These questions were and are a major reason why I have been a life-long professional philosopher. What should I do? Substantive ethics. Why should I do what I should do? Metaethics. What help does the discussion of the earlier chapters offer towards an answering of these questions? A philosopher looks at human beings, indeed!

1   The Status of Humans

4   What is mankind that you are mindful of them,
        human beings that you care for them?
5   You have made them a little lower than the angels
        and crowned them with glory and honor.
6   You made them rulers over the works of your hands;
        you put everything under their feet   (Psalm 8)

Well that tells it like it is, or at least, what the Psalmist –
King David – tells us it is. Let us not get too far ahead of
ourselves. Let everyone interested have a say on this
matter. In turn, I shall take the religious, those who
think that human status is given by the divine; then
the secular, those who think that human status is to be
found in the world; and, finally, those who think that
human status can and must be created by us, humans
themselves.

The Religious

Since we live in the West, start with the dominant religion,
Christianity. What I have to say applies more or less to the
other Abrahamic religions, Judaism and Islam. The Bible is
definitive and Genesis 1, read literally or metaphorically, is
explicit. There is a God, who is all powerful and all loving.
He is the Creator.

1 In the beginning God created the heavens and the earth.

2 Now the earth was formless and empty, darkness was over the surface of the deep, and the Spirit of God was hovering over the waters.

God set to work, making dry land and the seas and oceans. The sun too. Then plants: "11 Then God said, 'Let the land produce vegetation: seed-bearing plants and trees on the land that bear fruit with seed in it, according to their various kinds.' And it was so." Birds and fish and marine mammals: "20 And God said, 'Let the water teem with living creatures, and let birds fly above the earth across the vault of the sky.' 21 So God created the great creatures of the sea and every living thing with which the water teems and that moves about in it, according to their kinds, and every winged bird according to its kind. And God saw that it was good." On to land animals: "25 God made the wild animals according to their kinds, the livestock according to their kinds, and all the creatures that move along the ground according to their kinds. And God saw that it was good." Finally, to the climax, and it is a climax make no mistake.

26 Then God said, "Let us make mankind in our image, in our likeness, so that they may rule over the fish in the sea and the birds in the sky, over the livestock and all the wild animals, and over all the creatures that move along the ground."

27 So God created mankind in his own image, in the image of God he created them; male and female he created them.

God created plants and animals and so in themselves they are good. As we learn elsewhere, God cares about all His creatures. Matthew 6:26: "Look at the birds of the air; they do not sow or reap or store away in barns, and yet your heavenly Father feeds them." Matthew 10:29: "Are not two sparrows sold for a penny? Yet not one of them will fall to the ground outside your Father's care." But, as the ending of Matthew 6:26 reminds us, "Are you not much more valuable than they?" Plants and animals should know their place.

This is just the background for the whole drama that defines and creates the religion. We humans are made by God, so we are good. We are special, because we are made in the image of God. So, we have free will and responsibility. Then, straight away, we – Adam and Eve – spoiled it all by rank disobedience. We ate that wretched apple, the most unfortunate piece of fruit that a tree ever produced. Sinners, cast out of Eden, and worse, transmitting the sin to future generations – original sin. It is not that the newborn baby has sinned, but that, like all humans, it has a propensity to sin and, if given the opportunity, will sin. The greatest heroes of the Old Testament, the ones whom God loves above all others, are the greatest sinners. King David, so handsome, so brave, so talented. And then there is the lust for another man's wife, Bathsheba, and the dreadful act of putting the husband, Uriah the Hittite, into such a situation that inevitably he was going to be killed.

Fortunately, God did not give up on us and rectified the situation by coming down to Earth Himself, in the Form of Jesus, and offering Himself up as a blood sacrifice – only the death of God Himself would do the trick – thus making possible our eternal salvation. Other animals too? All Englishmen think that a heaven without dogs is an

7

oxymoron. If the Queen has corgis, can the Virgin Mary have less? Yet, within the Christian religion, and the same is true of Judaism and Islam, the central, favored status of human beings is a given. The same is true of other religions. Buddhism dates from the life and teaching of Gautama Buddha, born and living in Nepal around and after 550 BC. It is an atheistic religion, in the sense that it has no place for a Creator God, such as that of Christianity. Unlike Christianity, Buddhism is committed to the idea of reincarnation – that we have multiple lives in succession (samsara) – and actions and thoughts in this life can have implications for the life that we will live next. Ultimately the aim is to break out of this ongoing cycle of existences and achieve something called "nibbana" (also called "nirvana"). One is released from suffering – "dukkha" – and achieves a kind of state of non-being. This is not necessarily non-existence. We learn that it is endless and wholly radiant, the "further shore," the "island amidst the flood," the "cool cave of shelter" (no small thing given the Indian climate), the "highest bliss" (Harvey 1990, 63).

All of this takes place against the background of a rather complex ontology. There are an infinite number of universes, with galaxies, themselves clustered into thousand-fold groups. There are innumerable planets, and on them we find inhabitants, much like our planet and its denizens. Everything is subject to change, decay, and rebirth – often taking vast quantities of time (eons). Unlike Christianity, which has a beginning, a middle, and an end (the Second Coming), time seems like an endless string, going infinitely back and infinitely forward, and us somewhere hanging on in the middle. As we have this temporal dimension, so also we have other dimensions. It transpires that our level of existence is but one of five or six, and part of the process of rebirth is moving up or down these levels according to our behavior in this life. Right at the bottom is the hell-realm, "niraya," with vile beings tortured and subject to horrible nightmares. Then above this comes the level of "petas," ghostly creatures, somewhat akin to the phantom spirits of Western lore. The wilis (girls who die of heartbreak from being jilted) of the ballet *Giselle* would be eminently qualified here. Next up is the animal realm, obviously sharing space with humans, but in major respects lower forms of life. Humans come next and then above us are one or two levels for the gods – the "asuras," the lesser gods, and then the "devas," which include the "brahmas," the very highest form of being. Note, however, that everyone, at all levels of existence, is subject to life, death, and rebirth. Dukkha is omnipresent and the aim for all is nibbana.

It starts to seem that humans are special. We are above other forms of non-divine life, and one presumes that is the point of punishment or rehabilitation. If we behave badly, we are going to be reborn as a lesser form of life. Hitler has the prospect of many future lives as a codfish, in the oceans of Andromeda and like galaxies. There are beings above us, but then of course this is true of Christianity also. It has the angels, and these are as hierarchical as anything to be found in Buddhism. In the *Summa Theologica* of Aquinas (1265–74), for example, following tradition, he gives nine orders of angels, grouped in threes, ordered according to their closeness to God. Seraphim, Cherubim, and

Thrones; Dominations, Virtues, and Powers; Principalities, Archangels, and Angels. Traditionally this was put together in what was known as the "Great Chain of Being," an idea which goes back to Aristotle and his ranking of organisms in his *History of Animals* (Figure 1).

What is interesting about Buddhism is that, so convinced it is of the importance of humans, they can in respects perform at a higher level than the gods. In early Buddhism there is one major god, the Great Brahma. There are suggestions that he might have been the creator of the Earth. The Great Brahma himself encouraged such thinking. "I am Brahma, the Great Brahma ... the All-seeing, the Controller, the Lord, the Maker, the Creator ... these other beings are my creation." The Buddha, however, showed that the Great Brahma was mistaken. He was just a being like everyone else. Which has the interesting implication that, although the Great Brahma is a higher level of being than the Buddha, a human, it was the Buddha who was wiser and closer to nibbana. Compared with Christianity, Buddhism might not make such a show of humans being so very special, but it is right there at the heart of the religion.

The Secular

"The God of the Old Testament is arguably the most unpleasant character in all fiction: jealous and proud of it; a petty, unjust, unforgiving control-freak; a vindictive, bloodthirsty ethnic cleanser; a misogynistic, homophobic, racist, infanticidal, genocidal, filicidal, pestilential, megalomaniacal, sadomasochistic, capriciously malevolent bully" (Dawkins 2006, 1). No less than King David, the evolutionist

Figure 1 The Great Chain of Being from Ramon Lull's *Ladder of Ascent and Descent of the Mind*, 1305: God, angels, heaven, humans, beasts, plants, flame, rocks.

Richard Dawkins has a way with words. This is the famous – notorious – sentence that opens his *The God Delusion*, a work written with such passion that it could fit easily among the minor prophets of the Old Testament. Certainly, if anything is true, it is that, on the God question, no two people could be farther apart than Richard Dawkins and King David. Yet, it would be harder to find anyone more committed to the humans-are-special thesis than Richard Dawkins. "Directionalist common sense surely wins on the very long time scale: once there was only blue–green slime and now there are sharp-eyed metazoa" (Dawkins and Krebs 1979, 508).

He goes further, spelling things out. It is soon very clear that the sharpest eyed of the metazoa are human beings. Dawkins brings up the increasing employment by competing nations of ever more sophisticated computer technology. In the animal world, Dawkins sees the evolution of bigger and bigger brains. We won! Dawkins refers to a notion known as an animal's EQ, standing for "encephalization quotient" (Jerison 1973). This notion is a kind of cross-species measure of IQ, factoring out the amount of brain power needed simply to get an organism to function – whales require much bigger brains than shrews because they need more computing power to get their bigger bodies to function. With the surplus left over, one can then scale raw intelligence. Dawkins (1986) writes: "The fact that humans have an EQ of 7 and hippos an EQ of 0.3 may not literally mean that humans are 23 times as clever as hippos! But the EQ as measured is probably telling us something about how much 'computing power' an animal probably has in its head, over and above the irreducible amount of computing power needed for the routine running of its large or small body" (189). Even an organism with a low EQ probably does not need much help in making out the precise nature and import of that something.

Dawkins is not alone among his kind in seeing humans as top dogs, as one might say. Winners of the Crufts' Best in Show. Edward O. Wilson, of Harvard, myrme-cologist (ants) and sociobiologist, is the doyen of living evo-lutionists. In his major book *Sociobiology: The New Synthesis*, he thunders forth his position. Writing of social evolution, which is his focus, Wilson tells us that of all animals: "Four groups occupy pinnacles high above the others: the colonial invertebrates, the social insects, the nonhuman mammals, and man" (Wilson 1975, 379). He continues: "Human beings remain essentially vertebrate in their social structure. But they have carried it to a level of complexity so high as to constitute a distinct, fourth pinnacle of social evolution" (380). He concludes by speaking of humans as having "unique qualities of their own." He now launches at length into showing us how humans have crossed over and mounted the "fourth pinnacle" (382) – the "culminating mystery of all biology" (382). All this, as Wilson makes clear in subsequent writings, is very much part of the general picture. "The overall average across the history of life has moved from the simple and few to the more complex and numerous. During the past billion years, animals as a whole evolved upward in body size, feeding and defensive techniques, brain and behavioral com-plexity, social organization, and precision of environmental control – in each case farther from the nonliving state than their simpler antecedents did" (Wilson 1992, 187).

Cross the campus, from the hall of science to the halls of the humanities, and stop over in the Philosophy

Department. The great Greek philosopher Aristotle was neither Christian nor an evolutionist. Nevertheless, he knew where he stood on human beings. We may infer "that, after the birth of animals, plants exist for their sake, and that the other animals exist for the sake of man. . . . Now if nature makes nothing incomplete, and nothing in vain, the inference must be that she has made all animals for the sake of man" (Barnes 1984, 1256b15–22). Likewise, explaining why humans alone are bipedal: "of all living beings with which we are acquainted man alone partakes of the divine, or at any rate partakes of it in a fuller measure than the rest." Hence, "in him alone do the natural parts hold the natural position; his upper part being turned towards that which is upper in the universe. For, of all animals, man alone stands erect" (656a17–13). As always, status has its costs: "Of all female animals the female in man is the most richly supplied with blood, and of all animals the menstrual discharges are the most copious in women" (521a26–28). You have to take the wet with the dry.

Aristotle's thinking did not come from thin air. Go back to his teacher Plato, and turn to the *Timaeus*, where Plato talks of the design and creation of the universe by his version of the ultimate divinity, the Form of the Good, what in the *Timaeus* he calls the "Demiurge." He talks of the creation of humans.

> God gave the sovereign part of the human soul to be the divinity of each one, being that part which, as we say, dwells at the top of the body, inasmuch as we are a plant not of an earthly but of a heavenly growth, raises us from earth to our kindred who are in heaven. And in

> this we say truly; for the divine power suspended the head and root of us from that place where the generation of the soul first began, and thus made the whole body upright. (Cooper 1997, 90b)

Adding: "When a man is always occupied with the cravings of desire and ambition, and is eagerly striving to satisfy them, all his thoughts must be mortal." However: "he who has been earnest in the love of knowledge and of true wisdom, and has exercised his intellect more than any other part of him, must have thoughts immortal and divine, if he attain truth, and in so far as human nature is capable of sharing in immortality, he must altogether be immortal" (90c).

Obviously, although not Christian, Plato is working in far more of a religious than a secular context. Little surprise that the great Christian philosophers, notably St. Augustine, were able readily to interpret their Jewish-derived theology in the terms of Greek philosophy. However, we can see that Aristotle, who clearly owes much to Plato – all of the stuff about the divine being in the upper part of the body and hence humans walk upright – is starting to drain the divine out of the story. Aristotle believed in an Unmoved Mover, but his explanations are less dependent on the direct design of a benevolent being. As we come down through the centuries, more and more it became possible to push the God element back and out of the picture. Today, for instance, eminent English philosopher John Dupré argues in an entirely secular manner. But even more than the Greeks, if possible, his interest – his obsession – is humans and their status.

> Though I certainly don't accept that only humans are capable of thought, our forms of consciousness of which we are capable, are very different from those of other terrestrial animals. And human culture, though not unprecedented, involves the articulation and synchronization of a variety of roles and functions that is different in kind from anything else in our experience. (Dupré 2003, 75)

Having stated that he thinks we can genuinely speak of human freedom – something not within the scope of other organisms – Dupré concludes: "What is important for now is just to note that evolutionary continuity with the rest of life doesn't mean that there may not be features of human existence quite radically different from any found outside the human sphere" (75–76).

Dupré is not alone in this way of thinking. Other notable philosophers of today who stress the importance, the uniqueness, of human nature include the late philosopher of mind Jerry Fodor and the influential Thomas Nagel. I will not delve further into their claims, for here what interests me rather more is the fact that, just as you have Dawkins and Wilson endorsing a view of humankind central to their opponents, the religious, so here among secularists you have people on very different sides nevertheless endorsing a view of humankind central to their opponents. Among enthusiasts for Darwinian evolutionary theory today it would be hard to produce two names more readily than Richard Dawkins and Edward O. Wilson. Yet although the philosophers – Dupré, Fodor, and Nagel – are all evolutionists, they are no lovers of Darwinism. At the time of the celebrations (in 2009) marking the 200th anniversary of the birth of Charles Darwin, Dupré remarked somewhat sneeringly of enthusiasts as being tainted by "Darwinolatry" (2010). Fodor hopes for another paradigm. Apparently, "an appreciable number of perfectly reasonable biologists are coming to think that the theory of natural selection can no longer be taken for granted." Fortunately, "it's not out of the question that a scientific revolution – no less than a major revision of evolutionary theory – is in the offing" (Fodor 2007). And Nagel authored a book with the title *Mind and Cosmos: Why the Materialist Neo-Darwinian Conception of Nature Is Almost Certainly False* (2012). No comment. At least, no further comment, at this time. Something interesting is afoot; but leave it for now. We shall return to the matter. Here, the point being made is the extent to which human superiority is a conviction of people of very different conceptual backgrounds. Backgrounds that do not suggest agreement. But there is!

The Creationists

That's a bit of a joke, because the people with whom I want to conclude this chapter are about as far as possible from what we usually understand by "Creationists" – folk who take Genesis literally. Six-thousand-year earth history, six literal days of creation, humans last and made from mud, universal flood (Whitcomb and Morris 1961). My (small-c) creationists have little interest in religion generally, most especially not idiosyncratic, American, evangelical literalism. For me, they are creationists in the sense that they think

human nature, its worth and status, is created not discovered. They are certainly people who think humans
special. They would probably think you queer in the head
if you even asked such a question. But the specialness comes
in the fact that we have the ability to make ourselves special.
Other organisms do not have this ability.

A prime example of the kind of thinker I have in
mind is the French existentialist Jean-Paul Sartre. In his little
essay *Existentialism and Humanism* (1948) he writes:

> Existentialism is not so much an atheism in the sense that
> it would exhaust itself attempting to demonstrate the
> nonexistence of God; rather, it affirms that even if God
> were to exist, it would make no difference – that is our
> point of view. It is not that we believe that God exists, but
> we think that the real problem is not one of his existence;
> what man needs is to rediscover himself and to
> comprehend that nothing can save him from himself, not
> even valid proof of the existence of God. (5)

He explains what this means for humankind:

> My atheist existentialism ... declares that God does not
> exist, yet there is still a being in whom existence precedes
> essence, a being which exists before being defined by any
> concept, and this being is man or, as Heidegger puts it,
> human reality.
>
> That means that man first exists, encounters himself and
> emerges in the world, to be defined afterwards. Thus, there
> is no human nature, since there is no God to conceive it. It
> is man who conceives himself, who propels himself
> towards existence. Man becomes nothing other than what
> is actually done, not what he will want to be. (1)

I see the difference between this position and the two earlier positions as follows. For the religious, human uniqueness, human superiority, is something God-given, or in the case of a religion like Buddhism, as part of the cosmic order of things. God made us the way we are – intelligent moral beings – different from other animals, let alone plants. That is all there is to it. No further explanation is needed or possible. For the Buddhist, something comparable. That is the way of the cosmic order. That is all there is to it. No further explanation is needed or possible. For the secular, human uniqueness and superiority is something we find out there in the world. Take Wilson and his pinnacles. We have a degree of social order, of social complexity, of social functioning, that other organisms simply do not have. It is not a question of our thinking ourselves superior. We are superior! We can do things that other organisms simply cannot do. Likewise, for Dawkins. It is hard cheese on hippos, but they simply don't have the brain power – the computing power – that humans have. The Greeks, whether or not they are considered as more religious or more secular, are right in with all of this. The simple fact of the matter is that humans are bipedal. Warthogs are not. And the reason for this is that we have brains and the power of thought not possessed by warthogs. Moving down to the present, Dupré equally fits right in here: "our forms of consciousness of which we are capable, are very different from those of other terrestrial animals." Likewise with human culture. It "involves the articulation and synchronization of a variety of roles and functions that is different in kind from anything else in our experience." That's just the way that things are. Sorry!

I don't see the creationists, in the sense of the term as I am using it here, would disagree about our intelligence. Obviously, it is something that we have. But intelligence is only intelligence if you use it. If you veg out all day in front of the telly, you are not proving or creating your superiority. Cairn terriers can very happily do it a lot longer than you. We may well have societies way more complex than, let us say, the chimpanzees. More complex and efficient than the hymenoptera. If foraging ants get caught in the rain, they get washed away. If human shoppers get caught in the rain, they go to Starbucks and have a latte. But all of this is our creation. Not conferred by God. Not discovered in nature. We are, to use a popular phrase, "condemned to freedom." Simple as that. What you see is what you made. No more. No less. And if you decide that that makes us superior to all others, that is our judgment. Not something found in nature.

Thus, the three perspectives. I find all of this very intriguing. Very puzzling. In its way, quite exciting. I am sure that there must be more to the picture than this. Let's follow this intuition and see where it leads us.

2   Mechanism versus Organicism

Science?

What tool or tools do we have at our disposal? It is trite, although true and pertinent, to say we have our intelligence and our senses. But to what end should they be directed? Obviously, to understanding the world around us. Science! Is this going to be acceptable to devotees of the three approaches we encountered in Chapter 1? Insofar as the second group, the secularists, are concerned, science comes with the territory. They define themselves with it and against it as background. The third group, my creationists, really don't talk much about science. For Sartre, it is all about us. It is we humans who must define ourselves and create our own essence. That said, they can hardly be that indifferent to science and the way the world works. Suppose we were like ungulates, spending all day and every day seeking out and eating low-grade fodder. We are going to do a lot less essence making than as we are now. And if we want to push ourselves to greater heights, then science comes with the territory. Surely Mozart's *Le nozze di Figaro* is a triumph of humankind, and yet it hardly comes about without some knowledge of instrument building and opera house construction and much more.

Which then leads us to the first group, the religious. Staying now with Christianity, what says it about science?

Can science have any relevance to its claims? Start with what Christians say is the basis of their beliefs. Faith and reason. All stress faith; some include reason. Those of us who have taken only introductory philosophy classes might think that someone like St. Thomas Aquinas is all about reason. Those five proofs! But this is the kind of silly misreading when, for students, you pull out a couple of pages and ignore the literally hundreds of pages in which they are embedded. Aquinas not only endorsed faith, he put it ahead of reason. How else would the ignorant or lazy find out about God? "For certain things that are true about God wholly surpass the capability of human reason, for instance that God is three and one: while there are certain things to which even natural reason can attain, for instance that God is, that God is one, and others like these" (Aquinas 1259–65, 5). Reason – where we could be mistaken – is limited, and, in the end, faith – where we cannot be mistaken – is top dog. "The truth of the intelligible things of God is twofold, one to which the inquiry of reason can attain, the other which surpasses the whole range of human reason" (7). Limited perhaps, but for the person in this tradition, no less than for the secular thinker, science comes with the territory. Doing science, although obviously God senses in ways different from us, is what it means to be made in the image of God. Just as we have our moral task down here on Earth – to care for the sick and needy – so we have our intellectual task down here on Earth – to discover and understand His wonderful creation. "The heavens declare the glory of God; the skies proclaim the work of his hands" (Psalm 19:1).

Protestants tend to be a lot more iffy about reason. Notoriously, Martin Luther referred to "that whore reason." Unsurprisingly, there is a tradition, found for instance in the German Pietists, who put all the burden of God-belief on faith. Hugely important was the nineteenth-century theologian Søren Kierkegaard. He spoke of a leap of faith, meaning that religious commitment had to go beyond reason and evidence into the unjustified, the absurd. If it didn't, then in some sense faith is downgraded. You don't really need faith if reason is there, backing you up. "When someone is to leap he must certainly do it alone and also be alone in properly understanding that it is an impossibility ... the leap is the *decision*" (Kierkegaard 1992, 11–12, 102). And: "Faith is the objective uncertainty with the repulsion of the absurd, held fast in the passion of inwardness, which is the relation of inwardness intensified to its highest" (610). Karl Barth in the twentieth century followed in this tradition, as have many others. But, even apart from the natural theological nature of some claims in the Bible, like the just-quoted Psalm of David, this does not mean a disregarding of science. Science cannot be used to prove God or other faith claims, but it can certainly be used to delimit and clarify faith claims. You simply must not believe on faith or anything else that Noah's Flood was literally true.

What about the American literalists, the real Creationists? They really do seem anti-science. Sola scriptura. But note that this is not genuine, traditional Christianity. Augustine, four centuries after Jesus, was well into interpretation. "Words can in no sense express how God made and created heaven and earth and every creature that he created" (Augustine 1991, 88). American literalism is idiosyncratic evangelicalism of the early nineteenth century, formulated as a support for people in a young nation – especially those pushing the frontiers across the land – facing danger and poverty and hard work and loneliness and much more. With the coming of mechanized printing, the Bible was freely available to all and a helpful guide to behavior. Having trouble with the missus? "Wives, submit yourselves to your own husbands as you do to the Lord" (Ephesians 5:22). And so forth. Add to this that, when it suits them, literalists are into science as much as Richard Dawkins. The great natural history museum, the Field Museum of Chicago, is 300 miles north of the Creationist Museum in northern Kentucky. Both have explanatory displays of natural selection, and the Creationist display is as good as if not better than the one at the Field Museum. How come? Creationists today argue that Noah carried only "kinds," and after the Flood, back on dry land, these evolved quickly into the different species we see today. There was only one original Galápagos tortoise. Now, thanks to evolution, there are fifteen!

Root Metaphors

So, science it is. Before we go further, consider what that means. Popular is the thought that science is simply a matter of *Dragnet* writ large. Tell it like it is. "Just the facts, ma'am, just the facts." As we now realize, particularly since Thomas Kuhn's *The Structure of Scientific Revolutions* (1962), scientifc thinking is embedded in overall metaphysical world pictures – paradigms. Kuhn argues that paradigms are "incommensurable," and change from one to another is never particularly rational. While not unempathetic to this thinking, neither I nor anyone else thinks it the whole story. My *Darwinian Revolution* (1979) tried to show that Kuhn was right about the psychological aspects of theory change, but that reason and evidence were also important. Important here is that, in later writings, Kuhn identified paradigms with metaphors. This is a major insight. Science is deeply metaphorical – force, pressure, work, selection, arms race, Oedipus complex – and science overall is done within metaphors.

In the *Poetics*, Aristotle talked of them. "Metaphor consists in giving the thing a name that belongs to something else ... on the grounds of analogy" (Barnes 1984, 1457b). He tells us that metaphors have "clarity and sweetness and strangeness," and that "Metaphor most brings about learning." Although his agreement is not essential to my argument, I take Aristotle in some way to be arguing that metaphor is more than just heuristic – if we were grownup, we could speak literally without metaphor and we could find things out without metaphor – but, in some sense "creative": it is a non-eliminable part of understanding. An important point to which linguists draw attention is that metaphors often come in packages, with one metaphor holding together others (Lakoff and Johnson 1980). Think of argument as a battle. I went at him with hammer and tongs – full frontal assault. I gave way gracefully – retreat. I came back at him with a different point – new weapons. We agreed to disagree – armistice. And so forth. Even more importantly, some metaphors – root metaphors – embrace everything. In the words of linguist Stephen C. Pepper (1942): "A man desiring to understand the world looks about for a clue to its comprehension. He pitches upon some area of common sense fact and tries to understand other areas in terms of this one. The original area becomes his basic analogy or root metaphor." World pictures. Paradigms.

Most of this is reasonably non-controversial. Equally non-controversial is the fact that Western science has been governed by two root metaphors, and the Scientific Revolution, taking us from Copernicus at the beginning of the sixteenth century to Newton at the end of the seventeenth century was, above all, a change in those root metaphors (Ruse 2021). From the Greeks down to the Revolution, the world was seen through the lens of an organism. This was the ordering principle of understanding. Plato set the scene in his *Timaeus*, arguing that the Demiurge, identified with the Form of the Good, had designed the universe as an organism, and this is why we can and must think of it in terms of ends. The point or the purpose of the eye is to see and so analogously everything, organic and inorganic, must have an end or purpose. The world in a very real sense is living – Gaia-like, in today's terms. Aristotle, a biologist before he was a philosopher, dropped the external designer in favor of a more internally focused force; but, still, it meant that for true understanding one had to think organically, in terms of ends, or what Aristotle called "final causes."

Thanks to Augustine and Aquinas and others, this metaphor was readily given a Christian interpretation. The world itself was not and could not be God. It is His creation.

But as we humans show fully, God creates organisms, so why not the world as a whole? For all that he was one of the key figures in the Scientific Revolution, the astronomer Johannes Kepler was ever an enthusiastic Platonist. "The view that there is some soul of the whole universe, directing the motions of the stars, the generation of the elements, the conservation of living creatures and plants, and finally the mutual sympathy of things above and below, is defended from the Pythagorean beliefs by Timaeus of Locri in Plato" (*Harmonices Mundi*, 1619, in Kepler 1977, 358–59). Having given a Christian blessing to this kind of speculation, with an enthusiasm that might not have been totally appreciated either by the great Greek philosopher or by the preacher from Galilee, Kepler explored in some detail the analogies between the functioning of the Earth's soul and more famil-iar bodily workings, arguing that "as the body displays tears, mucus, and earwax, and also in places lymph from pustules on the face, so the Earth displays amber and bitumen; as the bladder pours out urine, so the mountains pour out rivers; as the body produces excrement of sulphurous odor and farts which can even be set on fire, so the Earth produces sulphur, subterranean fires, thunder, and lightning; and as blood is generated in the veins of an animate being, and with it sweat, which is thrust outside the body, so in the veins of the Earth are generated metals and fossils, and rainy vapor" (363–64).

But, by now, the organic metaphor was under threat. The world was changing. People were starting to use machines to aid them in daily life – clocks, notably, for telling time. And with this came needs of metals and related items. This did not always sit easily within the organic metaphor, the world as an organism. A metaphor, note, that applies to all objects living and non-living, something that worried the mineralogist Georg Agricola, 1495–1555. First, he gave the traditional argument.

> The earth does not conceal and remove from our eyes those things which are useful and necessary to mankind, but on the contrary, like a beneficent and kindly mother she yields in large abundance from her bounty and brings into the light of day the herbs, vegetables, grains, and fruits, and the trees. The minerals on the other hand she buries far beneath in the depth of the ground; therefore, they should not be sought. But they are dug out by wicked men who, as the poets say, "are the products of the Iron Age." (Agricola 1556, 6–7; the translation is by the future president of the United States, Herbert Hoover, who was a mining engineer at the time.)

Then Agricola went after the argument.

> If we remove metals from the service of man, all methods of protecting and sustaining health and more carefully preserving the course of life are done away with. If there were no metals, men would pass a horrible and wretched existence in the midst of wild beasts; they would return to the acorns and fruits and berries of the forest. They would feed upon the herbs and roots which they plucked up with their nails. They would dig out caves in which to lie down at night, and by day they would rove in the woods and plains at random like beasts, and inasmuch as this condition is utterly unworthy of humanity, with its splendid and glorious natural endowment, will anyone be so foolish or obstinate as not to allow that metals are necessary for food and clothing and that they tend to preserve life? (14)

In the modern age, rubbing in the point using the replacing metaphor, the organic metaphor was running out of steam. Quite apart from the fact that Copernicus, putting the sun at the center – heliocentric – was messing up Aristotle's cosmology which had the Earth at the center – geocentric (Kuhn 1957). Although Copernicus was a minor cleric, who died in good standing, he was going after the whole Christian world picture that saw the Earth as the place of the whole drama of humans, their creation, their fall, their salvation. And Kepler did not help with his discovery that the planets go not in circles, but in ellipses, thus destroying the happy thought (going back to Plato) that the heavens describe only the perfect figure, the circle. Galileo made matters worse, with his geocentric mechanics, and the story was made complete by Isaac Newton, who explained everything with his three laws of motion and the law of gravitational attraction, that made (supposedly perfect) heaven part of the same system as (supposedly imperfect) Earth.

Mechanism

A new root metaphor was called for: the world as a machine. That is the real import of the Scientific Revolution, a change of metaphors, from the organism to the machine. Now the world was conceived mechanically.

> At all times there used to be a strong tendency among physicists, particularly in England, to form as concrete a picture as possible of the physical reality behind the phenomena, the not directly perceptible cause of that which can be perceived by the senses; they were always

looking for hidden mechanisms, and in so doing
supposed, without being concerned about this assumption,
that these would be essentially the same kind as the simple
instruments which men had used from time immemorial
to relieve their work, so that a skillful mechanical engineer
would be able to imitate the real course of the events
taking place in the microcosm in a mechanical model on a
larger scale. (Dijksterhuis 1961, 497)

Expectedly, the French philosopher and mathematician René Descartes waded in enthusiastically, going to the
heart of the matter, as one might say, arguing explicitly that
the human body is a machine. In his *Discourse on Method*,
Descartes discussed Englishman William Harvey's work
showing that the heart is a pump (a machine) and the similar
mechanistic functioning of other bodily parts. "This will
hardly seem strange to those who know how many motions
can be produced in automata or machines which can be made
by human industry, although these automata employ very few
wheels and other parts in comparison with the large number
of bones, muscles, nerves, arteries, veins, and all the other
component parts of each animal" (Descartes 1637, 41).

In England, Robert Boyle likewise endorsed the
new metaphor, explicitly using a machine by analogy to
make his point.

[The world] is like a rare clock, such as may be that at
Strasbourg, where all things are so skillfully contrived
that the engine being once set a-moving, all things
proceed according to the artificer's first design, and the
motions of the little statues that at such hours perform
these or those motions do not require (like those of

puppets) the peculiar interposing of the artificer or any intelligent agent employed by him, but perform their functions on particular occasions by virtue of the general and primitive contrivance of the whole engine. (Boyle 1686, 12–13)

Now, particularly with an eye to our overall interests, let it be emphasized that the move from the organic model to the machine model was not a simple move from belief, Christianity, to non-belief, atheism. Key figures, starting with Copernicus, were believing and practicing Christians. Descartes, educated by Jesuits, lived and died within the Catholic Church. Boyle was a deeply committed Protestant. However, something was happening. With the organic metaphor, having God around is almost to be expected. With the machine metaphor, God is less prominent. He is needed to get things set up and working. But after that, in the words of one of the (already-quoted) preeminent historians of the Scientific Revolution, God became somewhat of a "retired engineer." Everything is just a matter of things in motion, eternally, governed by unbroken, unchangeable laws.

What one does therefore find, somewhat expectedly, is something of a move from theism – a personal God prepared to intervene in His creation – to deism, God as Unmoved Mover. Unsurprisingly, although Newton remained a communicant in the Church of England, theologically he moved towards denying the Trinity, towards Unitarianism. Jesus was not God incarnate. He was just another human being. Important, perhaps; human, certainly. At the beginning of the eighteenth century, more and more moved to this kind of thinking. We don't need the Bible and all of that revelation stuff. We can get all that we want from reason and nature, and if Jesus and company go by the wayside, tant pis. "Having proved that God requires nothing for his own sake, I shall now, the way being thus prepar'd, shew you, That the Religion of Nature is absolutely perfect and that external Revelation can neither add to nor take from its Perfection" (Tindal 1732, chapter 6).

Surely, this is a little bit too fast? God is not that absent. Machines have purposes, in the case of the clock telling time. Hence, world machines must have purposes, and, in ferreting these out, don't we come back to God? Ends, purposes, "final causes" – what more recent thinkers called "teleological" systems – continue to be an essential part of science. However, increasingly, people found that final-cause thinking is not that scientifically helpful. Take the moon. There are all sorts of interesting questions about why the moon keeps circling the Earth and doesn't fly off into space. Under the Aristotelian geocentric picture, the moon, like the sun and the planets, was kept in place by invisible crystal spheres forever circulating around the central Earth. Now this is all gone, and modern physics can take over. A machine explanation, powered by Newtonian gravitational attraction. Asking about the purpose of the moon seems not so useful. The moon exists in order to light the way home for drunken philosophers seems a joke rather than a serious claim. For all that, in the eighteenth century there was a club, the Lunar Society, that made full use of the heavens in this way, timing its meetings for the full moon.

Increasingly, final-cause talk was dropped. The English philosopher Francis Bacon (1605) likened them to vestal virgins – "the research into Final Causes, like a virgin dedicated to God, is barren and produces nothing." Descartes (1644) was no less contemptuous. How can we ever be truly certain as to God's intentions? We should not be so arrogant as to presume we can ferret out His ways and His ends. The world was seen as a machine, without purpose, stripped of value. It just is. Don't think this limitation or restriction unduly weakens the metaphor. As the poet Robert Frost (1931) said about the machine metaphor – pointing out that no one thinks the world had to have "a pedal for the foot, a lever for the hand, or a button for the finger" – "All metaphors break down somewhere. That is the beauty of it" (81). As Humpty Dumpty said to Alice about words, we are the masters of metaphor, not their slaves.

Mechanism Triumphant

Here we are. Two root metaphors and the one has pushed aside the other. The organic metaphor flourished for 2,000 years. Then it went. The machine metaphor took over. We, however, do not live in the sixteenth and seventeenth centuries. We live in the present. Did the story end three centuries ago and here we are now with things decided and unchanged? Ask first about the machine metaphor. No one could or should deny that the metaphor flourishes today, with triumph after triumph. The world is thought of as a machine. The task of science is to find out about it and how it works. As with machines, an important technique is going to be studying the parts and seeing how they go to make up the functioning of the whole. Reductionism. I am not going to make any general arguments to support my case. They are not necessary. But to show what I am talking about and why it makes sense – very good sense – let us turn to one of the triumphs of twentieth-century science, the discovery in 1953 of the double helix by James Watson and Francis Crick.

To prepare the way, look first briefly at one of the most important machines in the years preceding that discovery, the Enigma Machine used by the Nazis to encode their messages, thus protecting them from the prying eyes of the British and their allies. At the heart of the machine were a series of rotors that took information typed in – letters on the keyboard – and, mechanically moving around and making electrical connections, scrambled that incoming information (Figure 2(a)). Codes were changed daily, and you needed the codes to get the information back in an unscrambled form to be read. The machine was invented in the 1920s and taken up by the German military, which used it right through the war. As it happens, even by the early 1930s, Polish cryptanalysts were breaking through and finding ways to get into the machines and decode the messages. At the beginning of the Second World War, this information was passed on to the British, who thus were able to gather huge amounts of information about German intentions and movements. Important for us to note here is that, in order to understand how Enigma worked, the Poles and then the British had to take it to pieces and look at the parts individually, before finding out how they worked together (Figure 2(b)).

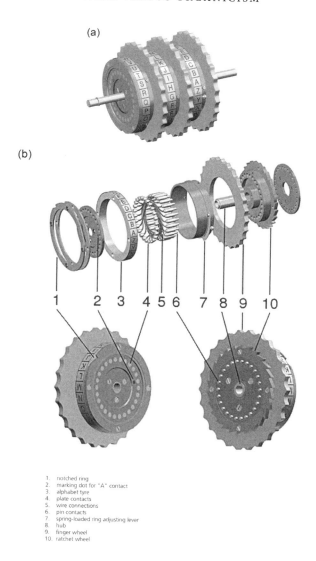

1. notched ring
2. marking dot for "A" contact
3. alphabet tyre
4. plate contacts
5. wire connections
6. pin contacts
7. spring-loaded ring adjusting lever
8. hub
9. finger wheel
10. ratchet wheel

Figure 2 Enigma: (a) rotor machine; (b) cogs.

35

A wonderful piece of machinery and an even more wonderful job of getting into it, and so the messages could be read by the Poles and British, finding how it worked. Now turn to the double helix. Again, it is all so famous there is not much time needed for pointing to the basic issues, or mechanisms as one might say. We have two (macro) molecules entwined around each other (Figure 3(a)). These are DNA molecules, deoxyribonucleic acid. It is they that carry the information that is needed to form the building blocks, proteins, that make up organisms. They are also the carriers of information from one generation to the next. Keeping the story simple, in complex (eukaryotic) cells, they are found on the chromosomes in the nucleus. To understand their working, as with Enigma you break things down into parts. A DNA molecule is essentially a chain, of linked parts, nucleotides. These come in four types: adenine (A), cytosine (C), guanine (G), and thymine (T) (Figure 3(b)). The way that these are ordered yields a "code," that can be (and was) deciphered, showing how the information is passed down the line. (You might have 1,000 triplets, codons, along a DNA molecule. Physically, the whole thing is about a millimeter long.) Another nucleic acid, RNA, ribonucleic acid, lines up against the DNA, copies the coded information, and then the RNA goes out into the cell and starts picking up amino acids, complex organic molecules. There are twenty such amino acids involved in the building process, which means that the genetic code (at its simplest) must group nucleotides in threes – as ACG, or TTC, or some such grouping. There are sixty-four possibilities, so there is redundancy, with some different groups picking up the same

(a)

DNA double helix

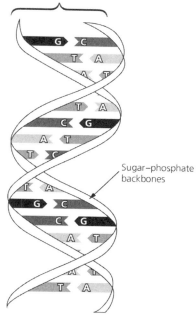

Sugar–phosphate
backbones

Adenine **(A)**   Thymine **(T)**   Guanine **(G)**   Cytosine **(C)**

(b)

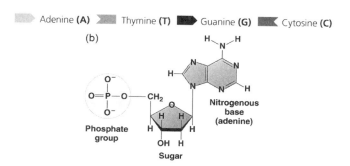

Figure 3 The double helix: (a) entwined chain molecules;
(b) nucleotide (adenine).

amino acids. RNA does rather more of its full share of the work, because it not only carries the information out from the DNA, but then gets to work making the triplets (codons) find their appropriate amino acids. As these get linked up a protein is produced – proteins are somewhat all-purpose molecules that not only supply material for structures but also regulate how things move along (enzymes).

Now I challenge you to see a great deal of difference between what humans do and what nature has wrought. The principles are exactly the same. You have intricate parts meshing together to get results. Of course, within the context of the machine, you can ask about purposes or functions. What does this particular cog do? Why this cog rather than some other? What does this molecule do? Why this molecule rather than some other? Why triplets in the code? Why not just pairs? (Because then you wouldn't have enough information to distinguish the twenty different kinds of amino acid.) But essentially you are looking at systems working blindly, according to unbreakable laws. If you want to come along and talk about ultimate purpose – transmitting information in such a way that the enemy cannot get at it, making it possible for beings to flourish and worship and thank their Creator – that is your business. But note it is value you ascribe there rather than value you find. Nature itself just is.

Organicism Redux

End of story? On to the next chapter? Not quite so fast! The machine metaphor was all very well when you were talking about planets and moons. Organisms qua organisms, however, seemed to be a different matter. Final-cause thinking didn't seem so irrelevant here. Robert Boyle knew the score. In his "Disquisition about the Final Causes of Natural Things," happily taking the opportunity to make a philosophical point while putting the boot into the French, he wrote: "For there are some things in nature so curiously contrived, and so exquisitely fitted for certain operations and uses, that it seems little less than blindness in him, that acknowledges, with the Cartesians, a most wise Author of things, not to conclude, that, though they may have been designed for other (and perhaps higher) uses, yet they were designed for this use" (Boyle 1688, 16). Boyle continued. Supposing that "a man's eyes were made by chance, argues, that they need have no relation to a designing agent; and the use, that a man makes of them, may be either casual too, or at least may be an effect of his knowledge, not of nature's." Apart from anything else, while this takes us from the opportunity to do science – the urge to dissect and to understand how the eye "is as exquisitely fitted to be an organ of sight, as the best artificer in the world could have framed a little engine, purposely and mainly designed for the use of seeing" (17) – it takes us away from the designing intelligence behind it.

Boyle was being forced into playing a double game here. His stance supposedly is not something threatening to the mechanical position. It complements it! How can this be so? Boyle is distinguishing between acknowledging the use of final causes qua science and the inference qua theology from final causes to a designing god. First: "In the bodies of animals it is oftentimes allowable for a naturalist, from the manifest and apposite uses of the parts, to collect some of the particular ends, to which nature destinated them. And in some cases we may, from the known natures, as well as from the structure, of the parts, ground probable conjectures (both affirmative and negative) about the particular offices of the parts" (Boyle 1688, 18). Then, the science finished, one can switch to theology: "It is rational, from the manifest fitness of some things to cosmical or animal ends or uses, to infer, that they were framed or ordained in reference thereunto by an intelligent and designing agent" (19). From a study in the realm of science, of what Boyle would call "contrivance," to an inference about design – or rather Design – in the realm of theology.

   Organisms are booted out of science into the realm of religion. A solution, but hardly a satisfactory solution, for all that, over the next century or more, some good biological science was done thanks to this uneasy compromise. Naturalistic mechanistic thinking in the physical sciences. Religion-entwined organismic thinking in the biological sciences. People went on worrying, notably the great, late-eighteenth-century, German philosopher Immanuel Kant. He turned to biology in the second half of the *Third Critique, The Critique of the Power of Judgment* (1790). Influenced by the biology of his day, Kant came right up against the problem of purpose, of final cause. In some sense, the parts of organisms are both cause and effect, with the kind of forward-looking, value-impregnated dimension that one expects in a world of purpose. The eye, for instance, brings about survival and reproduction, which in turn brings about another eye.

This seems to take us beyond the machine metaphor. "In a watch one part is the instrument for the motion of another, but one wheel is not the efficient cause for the production of the other: one part is certainly present for the sake of the other but not because of it. Hence the producing cause of the watch and its form is not contained in the nature (of this matter), but outside of it, in a being that can act in accordance with an idea of a whole that is possible through its causality" (Kant 1790, 32). Kant continues that it is a matter of organization or even self-organization. "This principle, or its definition, states: An organized product of nature is that in which everything is an end and reciprocally a means as well. Nothing in it is in vain, purposeless, or to be ascribed to a blind mechanism of nature" (33).

So where do we end up? Kant is driven to the conclusion that the teleology of biology is merely heuristic. It is a prop for needy humans.

> The concept of a thing as in itself a natural end is therefore not a constitutive concept of the understanding or of reason, but it can still be a regulative concept for the reflecting power of judgment, for guiding research into objects of this kind and thinking over their highest ground in accordance with a remote analogy with our own causality in accordance with ends; not, of course, for the sake of knowledge of nature or of its original ground, but rather for the sake of the very same practical faculty of reason in us in analogy with which we consider the cause of that purposiveness. (Kant 1790, 36)

Final-cause thinking is a guide, a heuristic. Kant could not say why, ultimately, we need it, but there we are. Although, that did not stop Kant from being rather nasty about biology. You want to make the life sciences equal to the physical sciences? No way! "[W]e can boldly say that it would be absurd for humans even to make such an attempt or to hope that there may yet arise a Newton who could make comprehensible even the generation of a blade of grass according to natural laws that no intention has ordered; rather, we must absolutely deny this insight to human beings" (37).

Romanticism

Truly, this is hardly more satisfactory than kicking the problem upstairs and letting God worry about it. It is no great surprise that there were those who simply gave up and pushed for a return to the organic model. Particularly around the end of the eighteenth century, beginning of the nineteenth century, especially in Germany, there grew up the Romantic movement – often in science designated *Naturphilosophie* – that called for a replacement of "the concept of mechanism" and a renewal of the organic metaphor, "elevating it to the chief principle for interpreting nature" (Richards 2003, xvii). Johann Wolfgang von Goethe, the poet, Friedrich Schelling, the philosopher, Lorenz Oken, the anatomist, are names often associated with the movement. Underlying it all was the philosophy of Plato, especially the Form of the Good, which integrates everything into one whole and gives underlying purpose to everything. The adolescent Schelling wrote a sixty-page essay on the *Timaeus*! The physical and the mental are not things apart. In an important way, the mental – the rational world of the

Forms (or ideas) – is the cause, certainly the informer of the physical – the changing world of experience.

Romanticism proved to be a hardy plant. Counter-movements like this, refusing to go with modern changes and thinking, sometimes show remarkable resilience. The Amish, for instance, accept very little of modern-day life, and yet they thrive. Interestingly, although the mechanists – or would-be mechanists like Kant – had little time for organic evolution – in the *Third Critique* Kant considered the idea only to reject it – there is reason to think that all the leading Romantics were evolutionists. Just as an organism grows fueled by its own internal causes, so Romantic evolutionism saw a kind of internal force or pressure driving the developmental process. Note, pushing away from the reductionism of mechanism, all is integrated and "holistic." The heart and the lungs and the liver work to the same end. So with evolution of organisms through time.

Like the coronavirus, Romanticism was highly contagious. In England, the perfect exemplar of this kind of thinking was Herbert Spencer (1852a, 1852b). He became an evolutionist early in the 1850s. Central to his thinking was the process named after the early-nineteenth-century French evolutionist Jean-Baptiste Lamarck – the inheritance of acquired characteristics. The giraffe stretches its neck as it reaches for leaves on the higher branches, and subsequent generations of giraffes are born with longer necks already in place. Spencer combined this with a somewhat idiosyncratic view of our reproductive powers, thinking that there is only a fixed quantity of seminal fluid. As organisms improve and (in the case of animals) brains grow, this fluid gets diverted from the loins and the end of reproduction to the production of ever-better organs of thought. Spencer was himself so far advanced that he never married and reproduced.

Lamarck proposed some kind of internal force, *Le pouvoir de la vie*, leading to ever greater complexity, at least in spirit and probably under the influence of the *Naturphilosophen*, especially Schelling (Rousset 1997, 393). In his *Histoire naturelle des animaux sans vertèbres* (1815), he wrote: "The rapid motion of fluids will etch canals between delicate tissues. Soon their flow will begin to vary, leading to the emergence of distinct organs. The fluids themselves, now more elaborate, will become more complex, engendering a greater variety of secretions and substances composing the organs." This was the position of Spencer, although typically he added some touches of his own. Not just complexity but integrated complexity – explicitly Spencer likened societies to organisms (1860) – and an odor of the second law of thermodynamics – Spencer believed that groups or societies would be in equilibrium, something disturbs them, and then they reachieve equilibrium at a higher level (Spencer 1864).

A remarkable gallimaufry of disparate ideas, grabbed from all over, put into the pot and presented as a supposedly coherent whole. For all the innovation, however, the thinking was deeply Romantic. The seminal influence on Spencer, via near-plagiaristic translations by the poet Samuel Coleridge, was the philosopher Friedrich Schelling. There was ever a reluctance by Spencer to acknowledge any influences upon his thinking, but even he admitted this much.

For the perception that there is a progress from a uniform to a multiform structure, and that this progress is the same in an individual organism and in a social organism, was a recognition of the progress from the homogeneous to the heterogeneous, though no such words were used. I had at that time no thought of any extension of the idea; but evidently there was the germ which was presently to develop. I should add that the acquaintance which I accidently made with Coleridge's essay on the Idea of Life, in which he set forth, as though it were his own, the notion of Schelling, that Life is the tendency to individuation, had a considerable effect. In this same chapter it is referred to as illustrated alike in the individuation of a living organism, and also in the individuation of a society as it progresses. (Duncan 1908, 541)

Like the Amish, the old root metaphor proved to have a long shelf life. The twentieth century saw an active if minority group of "organicists," some going back directly to the Romantics and others via Spencer (Peterson 2016). Henri Bergson, the hugely influential French philosopher at the beginning of the twentieth century, author of L'évolution créatrice, published in 1907 (English translation 1911), was the champion of the neo-Aristotelian life force, the élan vital – hence, better known as a "vitalist" rather than the more comprehensive "organicist." Yet, he was not that far out of the loop. The influences on Bergson are not hard to find. Jean-Gaspard-Félix Laché Ravaisson-Mollien, probably the most influential French philosopher of the second half of the nineteenth century, was a student of Schelling and a teacher of Bergson. Turning his gaze from East to West, it was nigh inevitable that, with such a pedigree, while being educated at the École Normale, Bergson became an enthusiastic Spencerian.

In the Anglophone world, in America in the early post-*Origin* era, the most influential biologist was the Swiss-born Louis Agassiz, transplanted to Harvard, the home of New England Transcendentalism, a philosophical movement with deep roots in German Romanticism – as with Spencer, gleaned mainly through the translations of Coleridge. Agassiz, who became a good friend of Ralph Waldo Emerson, had been a student in Munich. His teachers were Lorenz Oken and Friedrich Schelling. Of the latter, a classmate wrote: "A man can hardly hear twice in his life a course of lectures so powerful as those Schelling is now giving on the philosophy of revelation" (Lurie 1960, 51). In fact, Agassiz could never bring himself to accept evolution. All his students, including his own son Alexander, ignored their old prof's reservations and became enthusiastic evolutionists, of a Romantic kind.

The way was prepared for the arrival of the Englishman, Alfred North Whitehead, co-author with Bertrand Russell of the three-volume opus *Principia Mathematica*, in which they tried to show that mathematics follows deductively from the laws of logic. Coming to Harvard after the Great War, Whitehead switched from formal analysis to metaphysics. In 1846–47, making a triumphant entry into the New World, Agassiz gave a series of lectures at the Lowell Institute. Nearly 100 years later, in 1926, Whitehead did likewise. In his Lowell Lectures, published as *Science and the Modern World*, as with the Transcendentalists of the century before, Whitehead called for "the abandonment of the traditional scientific materialism, and the substitution of an alternative doctrine of organism" (99). Affirming: "Nature exhibits itself as exemplifying a philosophy of the evolution of organisms subject to determinate conditions" (115). The sciences are thus united: "Biology is the study of the larger organisms; whereas physics is the study of the smaller organisms" (129). Like Herbert Spencer, it is not easy to trace Whitehead's debts. They are there. "Whitehead's critique of scientific materialism and his philosophy of organism can be interpreted as efforts to develop Schelling's ideas more rigorously in the light of recent physics. For Whitehead, as for Schelling, nature is 'unconscious mind'" (Gare 2002, 36). Unsurprisingly: "Virtually every idea in science that inspired Whitehead was influenced in some way by Schelling's philosophy of nature." This was not chance. Like Spencer, he probably never read Schelling, but, like Spencer, just before his lectures he read a detailed exposition of Schelling's ideas. This combined with already-existing influences. "Schelling's evolutionary cosmology in which nature is seen as self-organizing ... formed the core of Herbert Spencer's evolutionary theory of nature, which then had a major influence on both Bergson and Whitehead." Organicism had staying power!

And now, with the way prepared, let us turn to today's preeminent scientific account of the arrival of humans and of their nature: Charles Darwin's theory of evolution through natural selection.

3   Darwinian Evolution

Human Prehistory

"Begin at the beginning," said the King of Hearts to the White Rabbit. That's good advice, so let's follow it. What is the beginning of a scientific story about human origins? Cover our options and play it safe. The Big Bang, which started everything, occurred about 13.8 billion years ago (Morison 2014). Whether there was anything before it or what caused it is a matter of speculation. The universe as we know it – the sun and the planets – is about 4.5 billion years old. Nothing lasts forever, and it is thought that the sun is about half-way through its lifetime. Not to worry. It isn't going to go out for a while yet. That means there will be ongoing solar energy for Planet Earth, which is thought to have been formed from detritus around the sun.

The causes of the origin of life are still in dispute (Bada and Lazcana 2009). In the sense of working according to established, unbroken laws, no one in the scientific world has any doubt that these causes are natural. The origins are findable, and one day perhaps soon will be found. Nobel Prizes for the winners! What we do know is that life seems to have appeared about as soon as it could have appeared, meaning as soon as the Earth and – especially – the water on its surface had cooled enough to allow life to flourish. You are not going to get much life if everything, everywhere, is 1,000°C. No one can be precisely accurate about these matters – Archbishop Ussher (1581–1656) pinned down the first day of Creation as Sunday, October 23, 4004 BC – but general opinion is that life started around 3.8 billion years ago (bya).

As we saw in the last chapter, it is macromolecules (strings) of nucleic acid that carry the information that builds organisms. At first, ribonucleic acid (RNA) was the key functioning unit. Later, deoxyribonucleic acid (DNA) came on the scene. As we also saw, we know a lot about how it functions and passes on the information. For the first half of life's history, the cells were simple – prokaryotes – without complex nuclei and other cell parts. Then, about 2 bya, some prokaryotes fused and more complex cells – eukaryotes – were formed. Complex life was off and running (Benton 2009). (See Figure 4.) The big event is generally thought to be the Cambrian explosion – about 550 million years ago (mya). This was when most of the major groups (technically known as "phyla") appeared – arthropods (insects), chordates (animals with a notochord, a kind of skeletal rod), mollusks (snails), and more. Because we humans have backbones, it is the chordates that matter to us. Evolution took us through our own particular sub-group (the vertebrates) from fish, to amphibia, to reptiles, to mammals and birds (Harari 2015; Reich 2018). We humans are mammals, our ancestors appearing about 225 mya, rat-like, nocturnal, and keeping well out of the way of those lumbering reptile brutes, the dinosaurs.

Mammals gave rise to the primates, about 50 mya or rather older, and now (from our perspective) the story starts

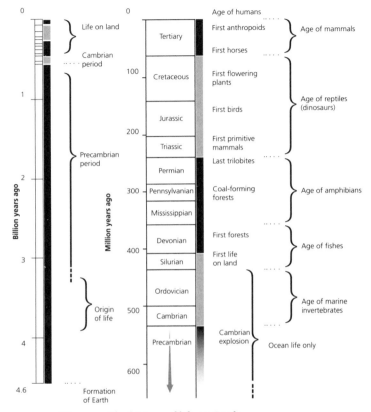

Figure 4 The history of life on Earth.

to get really interesting. First you get monkeys – although we are their descendants, the actual groups of animals (species) from which we come are now extinct – then the great apes, and finally the line that is going to lead to humans. Members of this line are known as "hominins." It is now thought, much to the surprise of many people about fifty years ago,

50

"Monkeys" and "apes"

New World Monkeys   Old World Monkeys   Gibbons   Orang-utans   Gorillas   Chimpan-zees   Humans

Figure 5 Humans' relationship to other primates.

we split off from the other great apes about 6 or 7 mya, and, even more to the surprise of many people about fifty years ago, our line split first from the gorillas and then from the chimpanzees. In other words, our closest great-ape relatives are the chimpanzees, and they are more closely related to us than they are to gorillas. (See Figure 5.)

As the late Stephen Jay Gould (1988) used to empha-size, the history of life is not like a poplar tree, pushing straight to the top. Leaving aside the complexifying fact that genes can be transferred across lines, in respects making a network a more appropriate metaphor, the history is much more like a bush than a tree. The line leading to us kept splitting and splitting. (See Figure 6.) Thanks to the insights of

(a)

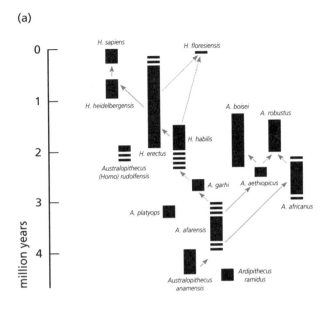

(b)

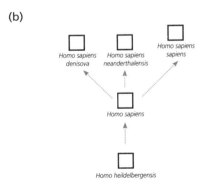

Figure 6 (a) Human phylogeny. (b) The human subspecies.

the eighteenth-century Swedish taxonomist Carolus Linnæus, biologists classify organisms in a hierarchical fashion, today thought to reflect histories (phylogenies). The lowest two levels are the genus (plural genera) and then below that the species. These are classes, but note they are classes of individuals. It is not the species as such which is the member of its genus, but the individuals of the species which are members of the genus. Species are given two latinized names, first the genus and then the particular species. Humans are *Homo sapiens*. If the species is then divided into subspecies, another name is added – as *Homo sapiens neanderthalensis*. Our human history (our phylogeny) of the past five or so million years opens with the *Ardipithecus* genus (5.6–4.4 mya). Then evolved from that – meaning the members of the later genus evolved from the members of the earlier genus – we have the Australopithecines (*Australopithecus*, 4.9–1.98 mya) – that's the one to which the famous fossil "Lucy" belonged (*Australopithecus afarensis*, 3.9–2.9 mya). (See Figure 7.) She (we know the remains are of a female) was discovered in Ethiopia in the early 1970s. What is remarkable – raises her to "missing link" status – is that she had a small brain, about the size of a chimpanzee as opposed to a human (this does not mean that she had a chimpanzee brain) – but, obviously, could walk upright. Not as well as today's humans; compensating, she was better at climbing trees. Probably Lucy was not literally our ancestor, but one of the group that included our ancestors (Johanson and Edey 1981).

Off to one side was the *Paranthropus* genus – and then finally the genus *Homo* (2.8 mya), our genus. This had several species, most famously first *Homo habilis*

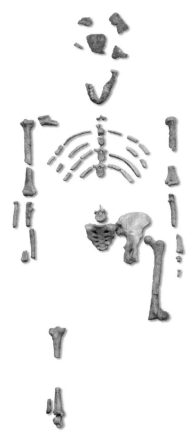

Figure 7 Lucy – *Australopithecus afarensis.*

(2.3–1.5 mya) and then *Homo erectus* (2–0.7 mya). Finally, *Homo sapiens* (0.35– mya), to which must be added the Neanderthals (0.4–0.04 mya), and the "hobbit," *Homo floresiensis* (0.19–0.05 mya). *Homo heidelbergensis* (0.7–0.3 mya) is part of the mix, but whether ancestral to all humans or just to the Neanderthals is a matter of debate. Now to be added is a kind of parallel group to the Neanderthals, the Denisovans. It could be that pre-humans moved out of Africa into Eurasia (more than a million years ago), where the separation of our ancestors from Neanderthals and Denisovans occurred. Then our line moved back to Africa (around 0.30 mya), and finally some of us went north (for a second time) to Eurasia (>0.10 mya), where we encountered long-separated Neanderthals and Denisovans (Reich 2018, 69). We know, through looking at the molecules in surviving fossils, ancient DNA studies, we Europeans are 2% Neanderthal. Five percent Denisovan in New Guinea. This is not true of today's Africans, who never lived side-by-side with the Neanderthals or Denisovans – that is, shared the same bed. Does this then mean that we are the same species as the Neanderthals and Denisovans? The definition of a species is a group of interbreeding organisms, reproductively isolated from other such groups. Despite the heroic efforts of generations of shepherds to prove otherwise, humans and sheep are different species. Clearly, we humans were never that isolated from Neanderthals or Denisovans. We were a different subspecies, but not a different species.

Darwinian Theory

Back to human pre-history. What brought all of this about? In one sentence: Charles Darwin's theory of evolution through natural selection. Separate two things. First, the *fact* of evolution, that all organisms, past and present, including humans, came by a natural developmental process from, as Darwin said, "one or a few forms." Second, the *causes* or *mechanism(s)* that brought this all about. Darwin first became convinced of the fact of evolution (Browne 1995). (See Figure 8.) This happened in March 1837, and then he spent eighteen months looking for an explanation, which he found at the end of September 1838. This search for a cause was not idiosyncratic. A graduate of the University of Cambridge, where unsurprisingly the work of the earlier Cantabridgian Isaac Newton was taken to be the model of good scientific practice, Darwin searched for the biological equivalent of the causal force of gravitational attraction. At the end of September 1838, he found it. Natural selection! After sitting on his ideas for some fifteen to twenty years – in 1842 he wrote a 35-page "Sketch" and in 1844 he wrote a 230-page "Essay" – spurred by the arrival of an essay with nigh-identical ideas by the naturalist Alfred Russel Wallace – he finally published his theory. *On the Origin of Species by*

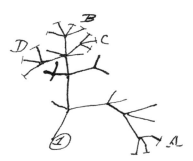

Figure 8 Darwin's first sketch of the tree of life in a notebook of 1837.

Means of Natural Selection, or the Preservation of Favoured Races in the Struggle for Life appeared towards the end of 1859.

Although it was accepting the fact of evolution that led Darwin to look for causes, in the *Origin* he started with the causes – following Newton who gave the laws of motion and of gravitational attraction – and only then moved on to the fact, the tree of life – following Newton who explained the heliocentric view of the universe. First, he introduced artificial selection, the picking and choosing that farmers and breeders do with animals and plants to get ever better specimens. He stressed how powerful this is, at the same time emphasizing that the changes are in the direction of properties – fatter cattle, shaggier sheep, more beautiful songbirds – the selectors want. There is an underlying program of design. To this end, one breeder said that selection "is the magician's wand, by means of which he may summon into life whatever form and mould he pleases." Another, speaking of the breeders of sheep, concluded: "It would seem as if they had chalked out upon a wall a form perfect in itself, and then had given it existence" (Darwin 1859, 31).

Now Darwin was off and running, showing how there is a natural equivalent of breeders' selection and that this too produces design-like organisms. Preparing the way, he argued that in all populations one finds variation – something needed for the building blocks of evolution. Although he had assumed it before, a decade-long study of barnacles had given empirical evidence of the ubiquity of this variation (Darwin 1851). Then come the key arguments. First, invoking the ideas of the late-eighteenth-century political scientist (and Anglican clergyman) Thomas Robert Malthus (1826), Darwin argued that in natural populations we are going to have a "struggle for existence." The potential rate of reproduction is geometric, whereas the potential rate of food increase is arithmetic, and the former – 1, 2, 4, 8, ... – outstrips the latter – 1, 2, 3, 4. "Hence, as more individuals are produced than can possibly survive, there must in every case be a struggle for existence, either one individual with another of the same species, or with the individuals of distinct species, or with the physical conditions of life." Adding: "Although some species may be now increasing, more or less rapidly, in numbers, all cannot do so, for the world would not hold them" (63–64).

Seizing on this conclusion, and invoking that variation for which he had argued, Darwin asked: "Can the principle of selection, which we have seen is so potent in the hands of man, apply in nature? I think we shall see that it can act most effectually." He further asked whether it can "be thought improbable, seeing that variations useful to man have undoubtedly occurred, that other variations useful in some way to each being in the great and complex battle of life, should sometimes occur in the course of thousands of generations?" He answered himself: "If such do occur, can we doubt (remembering that many more individuals are born than can possibly survive) that individuals having any advantage, however slight, over others, would have the best chance of surviving and of procreating their kind? On the other hand, we may feel sure that any variation in the least degree injurious would be rigidly destroyed. This preservation of favourable variations and the rejection of injurious variations, I call Natural Selection" (80–81). The action of natural selection over generations leads to wholesale change, evolution. What is crucial about the wholesale change brought on by natural selection is that, as in the domestic case, the change is not random. It is in the direction of design-like features – the hand and the eye – that will help their possessors in the struggle for existence. Adaptations! Thanks to natural selection, "we see beautiful adaptations everywhere and in every part of the organic world" (61).

Influenced by the different aims of breeders – utility versus beauty – Darwin introduced a secondary form of selection, sexual selection, where the struggle is for mates. Then Darwin added an important new component, the "division of labor," an idea going back to the eighteenth-century, Scottish economist Adam Smith (1776). Better avoid being a jack of all trades, and master of none. Specialization leads to speciation, as groups of organisms focus on particular tasks – one group eats one kind of plant and, avoiding competition, another group eats another kind of plant. Finally, Darwin was ready to introduce the fact of evolution, using the metaphor of a tree. "The affinities of all the beings of the same class have sometimes been repre-sented by a great tree. I believe this simile largely speaks the truth." Life grew upwards through time, with new branches representing new kinds of organism: "The limbs divided into great branches, and these into lesser and lesser branches, were themselves once, when the tree was small, budding twigs; and this connexion of the former and present buds by ramifying branches may well represent the classification of all extinct and living species in groups subordinate to groups." We arrive at the present. "As buds give rise by growth to fresh buds, and these, if vigorous, branch out and overtop on all sides many a feebler branch, so by generation I believe it has been with the great Tree of Life, which fills with its dead and broken branches the crust of the earth, and covers the surface with its ever branching and beautiful ramifications" (129–30).

The Consilience

The rest of the *Origin*, a good three-fifths, is devoted to showing how selection explains and is in turn confirmed by what we know of the nature and behavior of organisms in nature. (See Figure 9.) Starting this detailed overview – what Darwin's mentor at Cambridge, the philosopher William Whewell (1840), called a "consilience of inductions" – Darwin turned to social behavior as is exemplified by the hymenoptera – the ants, the bees, and the wasps. This was a subject of much interest in the nineteenth century, not least because so many people kept hives of bees. Why and how, for instance, do we find sterile workers in a nest, helping others but apparently going against the rule of adaptations always helping in the struggle for existence and reproduction? Darwin had no good grasp of genetics, but he offered a proto-version of what is now known as "kin selection" (Maynard Smith 1964). If you can help close relatives to reproduce – like the queen and your fertile sisters – then you are passing on your own features vicariously. Turning to the artificial-selection analogy, Darwin pointed out how we can pass on the desired features of the slaughtered steer by

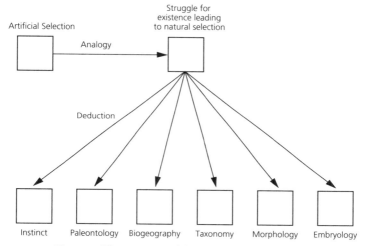

Figure 9 The structure of the *Origin of Species*. The deductions are little more than sketches.

returning to the breeding stock from which it came. If its fertile parents have fertile offspring, that will do the trick. Note that here, as always, Darwin thinks in terms of benefit to the individual, not the group. That goes right back to the beginning. Remember: "there must in every case be a struggle for existence, either one individual with another of the same species, or with the individuals of distinct species, or with the physical conditions of life" (63). In modern terms, Darwin was an ardent "individual selectionist" as opposed to "group selectionist."

After sociality, Darwin turned to paleontology and the fossil record, biogeography (the distribution of organisms around the globe), taxonomy, morphology, and embryology. All were shown to be understandable in the light of evolution through selection. Why are there fossils sharing the features of organisms quite different from each other today? Because those fossils are of shared ancestors. Why are the finches on different islands of the Galápagos archipelago in the Pacific so similar, but also different, and similar to, but again different from, the finches of mainland South America, the nearest land mass? Because ancestral finches came from the mainland to the islands, and then island-hopped and evolved apart in their isolation. One form has strong beaks, an adaptation for eating cactus and nuts, and another form has fine beaks, an adaptation for eating insects, and so forth. It was the realization that the different forms of finch were of different species that had first convinced Darwin of the fact of evolution. (See Figure 10.)

Why the taxonomic (Linnaean) hierarchy? As noted earlier, it reflects phylogeny (histories). Organisms in the same genus are more closely related to organisms in other genera and so forth. Anatomy. Corresponding to the question about the fossil record, there are the shared underlying forms between different species – underlying forms that Darwin's contemporary, the anatomist Richard Owen (1849), called "archetypes," likening them to the Forms of Plato. (See Figure 11.) Darwin explained the similarities – homologies – in terms of evolution through selection. Darwin was no fanatical pan-selectionist, insisting that every facet of the organic world had to be shown to be adaptive. Indeed, he had stressed this very point, while insisting that selection always comes first, and indirectly may explain the non-adaptive. "It is generally acknowledged that all organic beings have been formed on two great laws – Unity of Type,

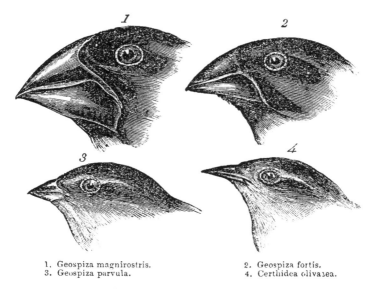

1. Geospiza magnirostris.
3. Geospiza parvula.

2. Geospiza fortis.
4. Certhidea olivacea.

Figure 10 "Darwin's finches" as they are now known. The different beaks are adapted for different food stuffs. The illustration is from Darwin's *Voyage of the Beagle* (second edition, 1845).

and the Conditions of Existence. By unity of type is meant that fundamental agreement in structure, which we see in organic beings of the same class, and which is quite independent of their habits of life. On my theory, unity of type is explained by unity of descent." Adding, however, that this means that "the law of the Conditions of Existence is the higher law; as it includes, through the inheritance of former adaptations, that of Unity of Type" (Darwin 1859, 206).

We come to embryology, by confession Darwin's own favorite. Why is the similarity of embryos of organisms

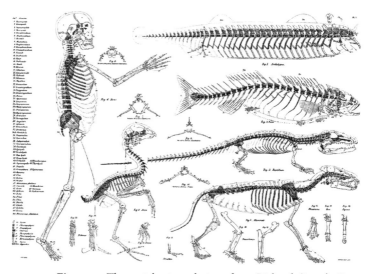

Figure 11 The vertebrate archetype from Richard Owen's *On the Nature of Limbs,* published ten years before the *Origin of Species.*

so very different as adults? Simply because, in the womb, conditions are similar for all and so there is no pressure on selection to separate them. Selection only kicks in as they move to adulthood. And so finally to the most famous passage in the whole history of science. Evolution, evolution through natural selection brought on by the struggle for existence, explains all.

> Thus, from the war of nature, from famine and death, the most exalted object which we are capable of conceiving, namely, the production of the higher animals, directly follows. There is grandeur in this view of life, with its several powers, having been originally breathed into a

few forms or into one; and that, whilst this planet has gone cycling on according to the fixed law of gravity, from so simple a beginning endless forms most beautiful and most wonderful have been, and are being, evolved. (490)

Humans

What about that most interesting of organisms, *Homo sapiens*? From the first, Darwin had always been stone-cold certain that we humans are part of the picture. In one of his private notebooks, written just after he had read Malthus and seized on what it all meant, we get the first explicit reference to natural selection – applied to humans and not just to humans but to our mental abilities. "An habitual action must some way affect the brain in a manner which can be transmitted. – This is analogous to a black-smith having children with strong arms. – The other principle of those children, which chance? produced with strong arms, outliving the weaker ones, may be applicable to the formation of instincts, independently of habits" (Darwin 1985, N 42). (Note that, like Spencer, Darwin was – and ever was – a believer in Lamarckian heredity, the inheritance of acquired characteristics.) In the *Origin*, however, fearing that too much discussion of humans would swamp everything, Darwin stayed away from the topic, dropping in only one provocative comment so people would not think he was being cowardly. "In the distant future I see open fields for far more important researches. Psychology will be based on a new foundation, that of the necessary acquirement of each mental power and capacity by gradation. Light will be thrown on the origin of man and his history" (488).

Darwin was right that it was the human question that did swamp all else – his theory became known as the "monkey theory" or "gorilla theory." The search for the missing link was on. Some claimed it was still around on the west coast of Ireland (Figure 12). Darwin, who had long been sick, hated this kind of controversy – he left it to his "bulldog," Thomas Henry Huxley – and probably he would never have written on humans were it not for the fact that Wallace took up spiritualism and began arguing that humans have features like hairlessness that could not have been produced by natural selection and so one must invoke occult causes (Browne 2002). Darwin was horrified, and at once set about writing and publishing an extension of his theory to our own species. *The Descent of Man and of Selection with Regard to Sex* appeared in 1871. Although there was much new material – of which more later – essentially it is more of the same. The one big difference, as flagged by the subtitle, is that now Darwin relied much more on his secondary mechanism, sexual selection, which is brought into play to explain such things as hairlessness. It leads to some very Victorian sentiments.

> Man is more courageous, pugnacious, and energetic than woman, and has a more inventive genius. His brain is absolutely larger, but whether relatively to the larger size of his body, in comparison with that of woman, has not, I believe been fully ascertained. In woman the face is rounder; the jaws and the base of the skull smaller; the

Figure 12 A cartoon of "Paddy" and his wife "Biddy," from the American humorous magazine *Puck* (1882). The legend is "King of A-shanti," being a pun on the name of an African tribe in what is today known as Ghana.

outlines of her body rounder, in parts more prominent; and her pelvis is broader than in man; but this latter character may perhaps be considered rather as a primary than a secondary sexual character. She comes to maturity at an earlier age than man. (Darwin 1871, 2, 316–17)

Darwin was a great revolutionary but no rebel.

After the *Origin*

For our purposes, the story of evolution from the *Origin* to the present can be told quickly. The big gap in Darwin's theorizing was an adequate theory of heredity. This did not arrive on the scene until the beginning of the twentieth century, when people discovered the experiments on pea plants and their interpretation by the monk Gregor Mendel, working in the Austro-Hungarian empire, back in the 1860s (Bowler 1989). Mendel's breakthrough was to see that the units of heredity, what we now call genes, are particulate – apart from occasional random changes (mutations) they are transmitted entire from generation to generation. At first, many thought Mendelism a challenge to Darwinism – take one or the other – but gradually they were seen as complementary, Darwinian selection working on Mendelian genes (Provine 1971). This was all brought together by the theoreticians around 1930 – in England, Ronald A. Fisher (1930) and J. B. S. Haldane (1932), and, in America, Sewall Wright (1931; 1932). A few years later the experimentalists and naturalists got into the act. In America, the Russian-born fruit-fly geneticist Theodosius Dobzhansky (1937), the German-born

ornithologist and systematist Ernst Mayr (1942), the pale-
ontologist George Gaylord Simpson (1944), and then,
bringing up the rear, the botanist G. Ledyard Stebbins
(1950). In Britain, the overall systematizer was Julian
Huxley (1942) – he was the grandson of Thomas Henry
and older brother of Aldous – and the chief naturalist was
E. B. Ford (1964) and his school of "ecological genetics."
The new theory or revitalized theory – known as "Neo-
Darwinism" in Britain and as the "synthetic theory" in
America – was off and running.

We shall learn much about this theory as this book
goes along. A couple of points and then we can move to
matters more philosophical. First, the modern theory is
completely and utterly Darwinian, in the sense that it is his
causal mechanism of natural selection – sexual selection too,
which today is often subsumed beneath the generic natural
selection – that is absolutely central. This said, in the
modern theory selection does have a rather different place
from that given to it in the *Origin*. There, it is up front and
center. We go from the population pressures to the struggle
for existence and from there to natural selection. Nothing is
rigidly formal, certainly not mathematical, but it seems that
(as always) Darwin was trying to be Newtonian in offering
laws (all populations have a tendency to increase in numbers
geometrically) from which one can deduce conclusions, like
the struggle and selection.

The modern theory is likewise Newtonian in this
respect – in the trade it is called being "hypothetico-deduc-
tive" – but much tighter and more formally mathematical.
What is crucial is that selection is deposed from its central

position (not role or importance) and the modern theory starts with genetics, Mendelian – more recently, as we shall see, molecular – thinking extended out from individuals to populations (Ruse 2006). The key hypothesis is the Hardy–Weinberg law. Genes are found on the chromosomes, generally in the heart of the cell, the nucleus, and because chromosomes come in pairs there are always two genes in an organism occupying the same paired position (locus). The Hardy–Weinberg law tells us that in any effectively infinite population with two forms of a gene (alleles, A_1 and A_2, say for blue or for brown eyes), in proportions let us say $p:q$ (where $p + q = 1$), then, if there are no disturbing factors, the ratio will remain at $p:q$ (in other words, the larger ratio will not swamp out the smaller ratio). Moreover, after one generation and for ever more the distribution will be $p^2 + 2pq + q^2$. The first term in this equation tells how many organisms will be homozygotes (having identical matching genes) for A_1, and hence have two A_1 genes, the second term tells us how many will be heterozygotes (having non-identical matching genes) and hence have one of each of A_1 and A_2, and the third tells us how many will be homozygotes for A_2, and hence have two A_2 genes.

The power of this law, which at first you might be inclined to dismiss as simply a bit of extended mathematics – Hardy was a pure mathematician with minimal biological knowledge – is that it provides a background of stability. Like Newton's first law of motion – if nothing happens, then nothing happens, bodies stay at rest or they continue in uniform motion – with the Hardy–Weinberg Law – if nothing happens then nothing happens, gene ratios stay the same – you now have a base to introduce disturbing causal factors. You know that what you introduce will have a quantifiable effect and will not simply vanish into a turbulent ocean. In the case of evolution, you can introduce the effects of migration into or out of a population, of the rate of spontaneous change of genes (mutations), and above all, natural selection. If, say, being a homozygote for A_1A_1 means you have no offspring whatsoever – in the language of evolution, its fitness is zero – you can start to work out how long A_1 will remain in the population and at what rate it will drop. Where things get interesting is where the heterozygote is fitter than either homozygote. A_1A_1 has no or some offspring, A_2A_2 has no or some offspring, A_1A_2 has more offspring. Then a fatal (or near-fatal) gene (in one or other homozygote) might stabilize in the population. The classic case of this is the sickle-cell gene. People with two copies of the mutant gene – homozygotes – usually die of anemia by the age of four. However, people with one copy of the gene – heterozygotes – have a natural immunity to malaria. In areas of Africa where mosquitoes mean malaria is a major threat, the sickle-cell gene persists, because the cost of the deaths of the sickle-cell homozygotes is balanced out by the successful lives of the heterozygotes (Allison 1954a, 1954b).

Molecular Biology – Friend or Foe?

The second point to be mentioned is the coming of molecular biology, the double helix of 1953. At first, the fear, particularly held by eminent evolutionists, was that Darwin's theory would be pushed to one side by the molecules. Fifteen years after Watson and Crick, we find Ernst Mayr writing that "enthusiastic but poorly informed physical scientists have lately tried very hard to squeeze all of biology into the strait jacket of a reductionist physical–chemical explanation." With the rhetorical flourish that characterized so much of his writing (and his whole personality), he concluded a critique of molecular intrusions into the evolutionary field by saying: "it is futile to argue whether reductionism is wrong or right. But this one can say, that it is heuristically a very poor approach. Contrary to the claims of its devotees, it rarely leads to new insights at higher levels of integration and is just about the worst conceivable approach to an understanding of complex systems. It is a vacuous method of explanation" (Mayr 1969, 128).

What is truly remarkable is that Mayr made this comment at a symposium also attended by his former student and future colleague Richard Lewontin, who just a year or two back had used molecular techniques to throw dazzling new light on one of the thorniest problems of traditional evolutionary thinking! In the 1950s, there had been sometimes bitter debate between Theodosius Dobzhansky (Lewontin's doctoral supervisor) and the Nobel Prize winner Hermann Muller. On the basis of such examples as the sickle-cell case – known as "balanced heterozygote fitness," because the heterozygote being the fittest meant that all pertinent alleles would remain at a steady level in the population – Dobzhansky argued that every natural population would have masses of variation, not just due to new mutations (as Darwin thought) but due to the action of natural selection. Masses of variation that were not going anywhere fast. Muller, to the contrary, argued that there is a basic norm, and selection quickly removes variants – unless, because of their superior fitness, in turn they become the norm (Lewontin 1974).

This was, for Dobzhansky certainly, no idle debate. A major criticism of Darwinism is that it is naïve to think that the right variations are going to come along just when they are needed. Variations, mutations, are random in the sense of not appearing according to need, and as like as not any new variation is going to be quite useless. With variation held in a population, there is always something that can be used. A new predator? If not a change in camouflage, then what about being distasteful, or an ability to hide, or simply an inclination to get the hell out of here? There is a range of possible responses and you will probably already have something that will do. It is like having to write an essay on dictators. If you must wait for the offerings of the Book of the Month Club, you will never pass the course. But with a library, if not a book on Hitler then how about a book on Stalin, and so forth.

The trouble was that, in the fifties, no one quite knew how to resolve the Dobzhansky–Muller debate. Muller was unmoved by Dobzhansky's worries, mainly because he was a geneticist rather than an evolutionary theorist and so didn't appreciate the worries. It didn't help that he was a Nobel Prize winner, and so knew by definition that what he said was right. Then, in the middle of the sixties, Lewontin and some others used molecular techniques to cut the Gordian knot. They showed how variations, which start ultimately at the molecular level, can be ferreted out. "Gel electrophoresis" uses electrical fields to show how variations caused by molecules of different sizes move at different speeds through a bed of gel (typically agarose made from seaweed). At once the Dobzhansky–Muller debate was over, with the former student showing that his old professor was right, there is much variation within natural populations. Mayr's worries were already dated. As an eminent Anglican theologian said at the end of the nineteenth century about the coming of Darwinism in an age of Christianity, it came disguised as a threat and proved to be a friend (Moore 1890). The almost comical – certainly ironic – thing is that molecular biology proved to be the handmaiden of evolutionary biology, not its master (or mistress). Evolutionary biology set the problems and molecular biology set about giving answers.

4   Mechanism and Human Nature

Homo sapiens

I want, in this chapter and the next, to look at the science of the last chapter, in the light of the root metaphors presented in the second chapter, with an eye to the three positions – religious, secular, creationist – of the first chapter. But first, let us dig a little more into what we know about human beings. Collating many studies of human origins, Daniel Lieberman (2013) writes:

> Hundreds of such studies using data from thousands of people concur that all living humans can trace their roots to a common ancestral population that lived in Africa about 300,000 to 200,000 years ago, and that a subset of humans disappeared out of Africa starting about 100,000 to 80,000 years ago. In other words, until very recently, all human beings were Africans. These studies also reveal that all human beings are descended from an alarmingly small number of ancestors. According to one calculation, everyone alive today derives from a population of fewer than 14,000 breeding individuals from sub-Saharan Africa, and the initial population that gave rise to all non-Africans was probably fewer than 3,000 people. (129)

One immediate implication of this is that you should expect that the genetic variation among humans is going to be at the low end of the variation one finds in species. Richard Lewontin is one who has seized on such facts as these and made it the premise for an argument concluding that there are no significant differences between races and that now the whole notion should be dropped. "Human racial classification is of no social value and is positively destructive of social and human relations. Since such racial classification is now seen to be of virtually no genetic or taxonomic significance either, no justification can be offered for its continuance" (1972, 397). However, as R. A. Fisher's student Anthony Edwards (2003) points out, this is a little quick. The overall variation may be small, but this does not mean that humans might not cluster in genetically delimited groups. Indeed, one much-cited study found just this (Rosenberg et al. 2002). The researchers looked at 1,056 individuals from 52 populations. They found that differences between groups are sufficiently strong that, if you run a cluster analysis across the large sample, you find that people sort into groups that correspond to ethnic sortings. Geographic Europeans come out as one genetic cluster and Africans come out as another genetic cluster. Pushing things, as you start to factor in more and more genetic information, the divisions get ever-finer and they continue to map ever-finer ethnic and geographical groups. For instance, the analysis picks out as anomalous a group in northern Pakistan. These are the somewhat isolated Kalash, who are believed not to be of the same ethnic background as the rest of their countrymen, but to have a European or Middle-Eastern origin. What can one say? "Genetic clusters often corresponded closely to predefined regional or population groups or to collections of geographically and linguistically similar populations" (2384). It seems there is no such thing as a standard human.

Not so quickly. Lewontin was not so far wrong about variation in the human species. The authors of the just-cited study reported: "Of 4199 alleles present more than once in the sample, 46.7% appeared in all major regions represented: Africa, Europe, the Middle East, Central/South Asia, East Asia, Oceania, and America. Only 7.4% of these 4199 alleles were exclusive to one region; region-specific alleles were usually rare, with a median relative frequency of 1.0% in their region of occurrence." Put things another way: "Within-population differences among individuals account for 93 to 95% of genetic variation; differences among major groups constitute only 3 to 5%" (Rosenberg et al. 2002, 2381). Other analyses back this up. Suppose, for instance, you wonder whether, as the Neanderthals and modern humans were different subspecies, we might today find different subspecies. Could the social differences represent different subspecies? Absolutely not! A recent careful study compared humans with chimpanzees (Templeton 2013). First the background theory.

> A race or subspecies requires a degree of genetic differentiation that is well above the level of genetic differences that exist among local populations. One commonly used threshold is that two populations with sharp boundaries are considered to be different races if 25% or more of the genetic variability that they collectively share is found as between population differences. (Smith et al. 1997)

Now, the facts on the ground. Comparison "confirms the reality of race in chimpanzees using the threshold definition, as 30.1% of the genetic variation is found in the among-race component, . . . In contrast to chimpanzees, the five major 'races' of humans account for only 4.3% of human genetic variation – well below the 25% threshold. The genetic variation in our species is overwhelmingly variation among individuals (93.2%)." No subspecies in the human species today!

Mentioned in the last chapter, a recent, hugely powerful, new technique for ferreting out information about the human past is ancient DNA (Reich 2018). We can extract DNA from old remains, early humans, Neanderthals, Denisovans, and this can be very revealing about our past. Everything discovered, especially about humans since we started to go it alone, tells us that populations have moved around the globe, displacing earlier groups and then in turn – by others, by the weather (ice ages) – being displaced. To suggest that we are fixed genetically and have been since time began is simply not true. We were always on the move. First, out of Africa and spreading east and west from the center of the present continent (less than 40,000 years ago, or 40 kya). Then the east moving west and taking over the whole continent (33–22 kya). The Ice Age comes and humans are squashed down into the Iberian peninsula, from whence they start to move north (19–14 kya). Finally spreading out from the Middle East (Turkey) across Europe again. We are not finished! The "Bell Beaker" culture, so named because of the distinctive nature of its drinking vessels, did a number on Britain. They moved in from Europe about 4 kya.

We are 90% descended from them and only 10% descended from those living there already. It was not our great grandparents who built Stonehenge. Our great grandparents sent the builders on their way (to extinction) (Reich 2018, 106–7). Non-stop churning, placing and replacing.

Final Causes

From now on, I am going to assume that intra-group variation is not important. What, then, about Darwinian evolutionary theory and its relevance to the human question? One thing that you can say is that Darwin's theory is as teleological – final-cause impregnated – as anything in Aristotle. So that's going to be a factor. The question is: in what sense? In the discussion of embryology in the *Origin*, Darwin asks why in some species – as opposed to others like the butterflies – the young closely resemble the adults.

> In certain cases the successive steps of variation might supervene, from causes of which we are wholly ignorant, at a very early period of life, or each step might be inherited at an earlier period than that at which it first appeared. In either case (as with the short-faced tumbler) the young or embryo would closely resemble the mature parent-form. We have seen that this is the rule of development in certain whole groups of animals, as with cuttle-fish and spiders, and with a few members of the great class of insects, as with Aphis. With respect to the *final cause* of the young in these cases not undergoing any metamorphosis, or closely resembling their parents from their earliest age, we can see that this would result from the two following contingencies; firstly, from the

young, during a course of modification carried on for many generations, having to provide for their own wants at a very early stage of development, and secondly, from their following exactly the same habits of life with their parents; for in this case, it would be indispensable for the existence of the species, that the child should be modified at a very early age in the same manner with its parents, in accordance with their similar habits. (Darwin 1859, 447–48, my italics)

Darwin is interested in efficient causes, in this case presumably natural selection, but he is as interested in final causes: why does it happen? Get the question right. Darwin did not deny teleology. He talked happily of final causes, meaning that it makes good sense to ask the purpose of features of embryological development, in this case of the earliest stages looking like the later stages. But this lack of change is not a product of an external designer or of internal forces. Nor is the teleology purely heuristic. Organic features are products of natural selection. Uniformity in development helped the possessor in the past, so we assume it will help the possessor in the future. We could be wrong, but that is how we think, and how we will continue to think unless and until changed circumstances introduce new selective pressures.

Darwin is hugely important. The New Atheists think it was because Darwin refuted God. Chicago evolutionist Jerry Coyne (2015) tells us that Darwin's theory is the "greatest scripture killer" ever (xii). This is completely to miss the point. Darwin is important because he completed the Scientific Revolution (Ruse 2017, 2021). Final causes can be subsumed under the machine metaphor. The world is to be seen mechanistically as well as from a reductionist perspective. We don't need to invoke a Designer, as with Plato. We don't need to get into neo-Aristotelian vital forces like the *élan vital*. The teleology is indeed heuristic. Those funny plates along the back of the stegosaurus have given rise to a cottage industry of would-be explanations. (Generally accepted today is that they were for heat control of the cold-blooded animal – catching the sun's rays in the morning, and then acting as cooling fins in the heat of the day.) But the teleology is more than heuristic. Organisms really are design-like. It is not a weakness that we see this or want to do the impossible and explain it. Darwin does explain it. Adaptations worked in the past. We expect them to work in the future. We may be wrong. Our sweet tooth today is hardly of much value. In fact, if you look at the average American today, it is a real problem. But that is the way of the world. In the past, it was of value to our ancestors to crave sweet things and to search out sources, like the honey in the nests of bees. In short, as with all genuinely mechanical scientific theories, Darwin's theory allows for relative value, but there is no absolute value. That has been drained from our understanding of the world. All we have are laws in motion.

Intelligence

Now, with an eye to the issues raised in Chapter 1, push the discussion of Darwin's theory a little. For the Christian, and truly for the others also, what makes humans special is

81

our intelligence and ability to find out about things – epistemology – and our moral sense, judging right and wrong – ethics. Leave the second for now. Does Darwinism have anything to say about humans as intelligent, thinking beings? It certainly does. But let's not do all of the work on our own. See if we can get some guidance from the philosophers. Not all philosophers. Our focus is on Darwinism; we need to turn to those who take seriously naturalism in some sense – willing to think our intelligence is a function of us as human beings and not simply something conferred from on high. Two names come at once to mind, David Hume and Immanuel Kant. These were philosophers who saw that the mind had to be contributing actively to knowledge, not just passively processing that which it receives. Using the modern metaphor, if the brain is a computer, the hardware, then the mind is the software, and it cannot just wait until something is fed in by the computer's user and then start from scratch. It must be pre-programed to some extent. David Hume, in his celebrated analysis of causation, saw this. Thinking causally is obviously important to human beings. Why throw a spear at an animal if you don't think it is going to go in the air towards its target? Why knock two stones together if you don't think you will get a spark? Why spend endless years of your young life if you don't think you will get a doctorate at the end of it? Hume ended by realizing that the notion of causality is not given to us in experience, but something we must furnish in our minds. A happens; B follows. It is we who make the connection, saying that A caused B. He writes: "from constant conjunction, objects acquire a union in the imagination. When the impression of one becomes present to us, we immediately form an idea of whatever usually accompanies it; and consequently we can lay this down as one part of the definition of opinion or belief, that it is an idea related to or associated with a present impression" (Hume 1739–40, 65).

"Objects acquire a union in the imagination." Let's agree that this is so and that the mind does furnish the causal connection, or rather the conviction of causal connection. But, with respect, so what? Why should any of this matter? If it is just weakness or imagination, then that doesn't bode well for understanding. Truth or knowledge seem very distant. It is here that Immanuel Kant steps forward, taking Hume's fundamental insight and trying to answer the philosophical questions about truth. Hence, his notion of the synthetic *a priori*, something stemming from the fact that the mind is set up to process information and produce knowledge.

> Empirical judgments, so far as they have objective validity, are judgments of experience; but those which are only subjectively valid, I name mere judgments of perception. The latter require no pure concept of the understanding, but only the logical connection of perception in a thinking subject. But the former always require, besides the representation of the sensuous intuition, particular concepts originally begotten in the understanding, which produce the objective validity of the judgment of experience. (Kant 1783, 298)

We get information from without, we process it, and then we can navigate and relate to reality. Because the information comes from outside, we must process it. It is not given to us directly. It is this processing that makes our understanding synthetic. However, this is not the whole story. Something is given by the mind. In this sense, we are dealing with the *a priori*.

But why necessary? Kant's answer is that these are the conditions – what Kant calls "categories of the understanding" – for thinking rationally at all! If you don't think this way, nothing works. "All empirical knowledge of objects has to conform to our a priori concepts, because if it doesn't then nothing is possible as an object of experience. And that is how matters stand. All experience contains, in addition to the intuition of the senses through which something is given, a concept of an object that is given in intuition (i.e. that appears)." Continuing: "Thus, concepts of objects as such underlie all experiential knowledge, as a priori conditions that it has to satisfy; so the objective validity of the categories as a priori concepts rests on the fact that it's only through them that experience is possible. . . . Since it is only by means of them that any object of experience can be thought at all, it follows that they apply necessarily and a priori to objects of experience" (Kant 1787, 126).

We are starting to get close, but there is still a move that is needed. How does this whole scheme come about? Could it be that the synthetic *a priori* setup just is? Kant was the child of Pietists, so one very much suspects that it is God who is doing the heavy lifting. However, it is a cardinal rule of Kantian philosophy that God be kept out of the discussion. Again, one suspects Pietism, because for the Pietist it is faith not reason that is the basis of religion. Reason is just fine in its proper place, but it is not in the God-business. We are stuck. All we know is that – "If you don't think this way, nothing works." And that leads to Darwin.

As soon as he grasped natural selection as the cause of evolution, Darwin was applying it to humans. He was a scientist not a philosopher, but he had had a good classical education, so he at once thought in those terms when trying to see just how evolution would affect our thinking. "Plato Erasmus says in Phaedo that our 'necessary ideas' arise from the preexistence of the soul, are not derivable from experience. – Read monkeys for preexistence" (Darwin 1987, M 128; the Erasmus referred to here is Darwin's older brother). This is the key to human knowledge. Our brains, to use the modern analogy, are computers but they are programed by natural selection to think in certain ways rather than other ways. Buy into the Kantian program. We think in certain ways rather than other ways because of the *a priori* categories of the understanding. These are in the mind before we start. However, they are not really *a priori*. They are Darwinian adaptations, because those of our would-be ancestors who had them – had them, not by design but by the regular processes of selection on random variations – survived and reproduced, and those of our would-be ancestors who did not have them, did not survive and reproduce. What's the point of thinking as we do rather than some other way? What's the final cause of our thought processes? Success in the struggle for existence.

Simple as that! It is so obvious when you think about it. Not unanswerable facts of existence, not God, just success and failure. Take mathematics, the prime example of the synthetic *a priori*, being in some sense constructions out of relations of space and time. Those proto-humans who saw two bears go into a cave and only one come out and who then decided to sleep somewhere else have successors today. Those proto-humans who saw two bears go into a cave and only one come out and who then said "Let's get out of the rain" are not around to tell us what happened when they went into the cave. Take causality. We have encountered the consilience of inductions. It is vital in science, but it is not confined to science. The detective looks for the blood stains and the method of killing and the opportunity and the motive and the broken alibi and declares – "Professor, I accuse you of murdering your junior colleague, because she was about to reveal that you have not been contributing your share to the coffee fund." Same back then. At the end of a long day hunting and gathering, you finally get to the pool where you can have a nice long cool drink. You see paw-prints in the mud. The bushes are trampled on. There is a sound of growling in the undergrowth. You say – "I'm not really that thirsty" – and keep on moving quickly. You will live to hunt and gather another day. You say – "Tigers? Just a theory not a fact. Nothing to worry about." Your hunting and gathering days are over.

Darwin's theory of evolution through natural selection shows us all how it works. Of course, it doesn't on its own show us how to solve Fermat's last theorem. It does give us the tools to solve it. It also shows us that a lot of the discussion about nature versus nurture is otiose. Our understanding, our knowledge, is so clearly a function or product of both. A child is born and raised by its parents.

Day in and day out, it is being fed information. About eating, about walking, about speaking, about not getting too close to the fire, about being careful around dogs, and so much more. Nurture is vital. But nurture must have something on which it can work. My cairn terriers cannot walk upright or speak a language. Goodness only knows what they would think and do if they walked into a classroom and a junior professor were lying there, on the floor, with a knife through her heart. Wag their tails, probably. It is a two-way thing. Those proto-humans who did not have the selection-given categories were in trouble. Those proto-humans who took no interest in the growth and upbringing of their children were in trouble.

Pragmatism

I am not writing a history of philosophy, but any professional philosopher knows that this kind of thinking is the essence of the whole Pragmatist approach to knowledge. Kantian categories, naturalized by Charles Darwin's theory of evolution through natural selection. "[New England] Cambridge pragmatism was, and is, more indebted to Kant than to any other single philosopher" (Murphey 1968, 9). Look at the early thinking of the greatest of them all, Charles Sanders Peirce. This comes in a series of articles published in *Popular Science Monthly* between November 1877 and August 1878. For Peirce, a key notion is that of a "habit." "That which determines us, from given premises, to draw one inference rather than another, is some habit of mind, whether it be constitutional or acquired"

(Peirce 1877, 3). I read habits as being up-to-date versions of Kant's categories of the understanding. Take causation and think of how we determine causation through consilience. For Kant, in some way, our mind is predetermined to guide us into certain sorts of thinking rather than others. Beware tigers, rather than tigers just a hypothesis. Naturalized, these come as habits. You think or reason this way rather than another way. But what determines the choice? Not some kind of transcendental necessary conditions of thinking. Rather, the practical consequences. "The essence of belief is the establishment of a habit; and different beliefs are distinguished by the different modes of action to which they give rise." Spelling this out: "Consider what effects, that might conceivably have practical bearings, we conceive the object of our conception to have. Then, our conception of these effects is the whole of our conception of the object" (Peirce 1958, 5.402). Note, however, it is not the consequences that are true or false. They happen or don't happen. It is the belief, informed by habit, that is true or false. It is the consequences that determine what is true or false. What is important is that this is a group effort: "reality is independent, not necessarily of thought in general, but only of what you or I or any finite number of men may think about it" (Peirce 1878).

More recent thinkers in the Pragmatist tradition make very similar points. Significantly, the best-known of them all, Richard Rorty, is adamant that we humans are at one with the rest of the living world, and that this is a consequence of being a Darwinian. "Darwinism requires that we think of what we do and are as continuous with what amoebas, spiders, and squirrels do and are" (Rorty 1998, 295). And he is openly a Pragmatist about all of this, with Dewey as his hero rather than Peirce, who is mine. He writes that Pragmatists "should see themselves as working at the interface between the common sense of their community, a common sense much influenced by Greek metaphysics and monotheism, and the startlingly counterintuitive self-image sketched by Darwin, and partially filled in by Dewey" (41). Couldn't have said it better myself!

Relativism?

Now before we – I – get too satisfied and declare victory and go out for a beer, hold for a moment and ask how the three groups of Chapter 1 might react to all of this. My suspicion is that the first two groups, the religious and the secular, might have a real worry about relativism. They both see humans as beings of real intellectual power and integrity. Beings capable of disinterested inquiry into reality, no matter where the search leads. But, with all the emphasis on pragmatic approaches and solutions, is that possible? Is that guaranteed? If you look at human societies today, and in the past, there is not much cause for optimism. Think of the false beliefs that people have had about the nature of the world, and the vile practices that they have endorsed. And yet, some of these societies have done very well. The American South in the years before the Civil War was thriving. A great deal better than the American South today. To make everything pragmatic is to assume too readily that everyone shares the ideals of an educated Englishman – "better Socrates dissatisfied than a fool satisfied" sort of thing. (And note the nationality of the chap who said that!)

We can speak part way to these very proper worries. No one could deny that there is cultural relativism with respect to what we believe. In the realm of science, of evolutionary studies, if we did not come from a Judeo-Christian background with the obsession about origins, I am not sure we would have an evolutionary perspective of the kind we have today – the subject of the last chapter. I am not saying we would all be biblical literalists – that is our way of thinking and wrong – but we might cut the pie up differently. Processes rather than origins sort of thing. In a Popperian fashion, you might think the best theory will triumph – ours! I am not quite so sure. Either way, I don't think this detracts from the dignity of the human intellect. Plato, Aristotle, Kant – none of them were evolutionists – and I am not about to tear up their entry tickets to the halls of the elite.

Moving from the cultural to the biological, you can see now why I wanted to start this chapter with more empirical information about our own species. There is simply no reason to think that the biology of *Homo sapiens* points to relativism at the intellectual level. If we go intergalactic, perhaps things are a bit different. Suppose, rather than through sight, we got around using pheromones and perhaps echo location. I don't see us having much of a theory of optics. We might not have the same mathematics as we have today, but it couldn't be that different. It couldn't say that $2 + 2 = 5$. That's our mathematics and it's wrong. It is the same with causation. We pull back from the fire because we think fire causes burning and burning causes pain. Natural selection clearly had a hand in all of this. I suppose it is possible to have some other system. We think that fire is pure and we are unclean and so the deity punishes us if we get too close. But there must be some kind of underlying sense of causality, whatever you think it is. I may be all wrong about this. I will confess to empathy with the population geneticist J. B. S. Haldane (1927), who maintained that the world is not only queerer than we think it is, but queerer than we could think it is. This is a very Darwinian comment. As Richard Dawkins (2003) points out, our evolved faculties look after us rather well. Yet, there is no reason to think that adaptations to get out of the trees and onto the plains are going to fit us to solve everything. Until they break down, I am not going to worry too much about relativism.

Humans Special?

Back to our three groups. Irrespective of whether the optimism of the theologian (Aubrey Moore) about the friendliness of Darwinism is justified, he is certainly closer to the truth than Coyne and his scripture killing. Darwin started life as a conventional Anglican, became a deist in his early twenties, and this lasted most of his life right through the writing of the *Origin*. This theological stance was an important component in his becoming an evolutionist. Deism sees a world of unbroken law, with an Unmoved Mover behind it. Evolution is support for this position, rather than refutation. Towards the end of his life, Darwin became an agnostic. Like many late-nineteenth-century former Christians, it was theology not science that drove him away from any form of God-belief. Darwin could not stomach the fact that his father and his brother, two of the finest men he had ever known, would go to hell because of their non-belief. Like almost all those former Christians, Darwin was never an atheist. Indeed, in the *Origin* itself, in a passage that he never changed through all six editions (the last in 1872), Darwin openly avowed empathy for deism. "To my mind it accords better with what we know of the laws impressed on matter by the Creator, that the production and extinction of the past and present inhabitants of the world should have been due to secondary causes, like those determining the birth and death of the individual" (Darwin 1859, 488).

Christian metaphors and tropes abound in the *Origin*. The tree of life for example. Thanks to the Elizabethan Settlement of the late sixteenth century, steering a path between the proselytizing of the martyrdom-intent Jesuits and the joy-destroying gloom and doom of the Calvinists – newly returned from Geneva where they sat out the reign of Bloody Mary (1553–58) – the English turned in a big way to natural theology, which fit nicely with the growing industrialism of the country. No surprise then that the really crucial Christian influence was, thanks to Darwin's university training, the thought of the Reverend Archdeacon William Paley, author of numerous textbooks, including *Natural Theology* (1802), where he argued that the existence of God is shown by the design-like nature of organisms. The eye is like a telescope, the telescope had a designer, hence the eye had a designer – the Great Optician in the Sky. It is absurd to deny this. Darwin used to joke that he could have written all of Paley's arguments by heart; except, it wasn't a joke (Darwin 1958). As we have seen, the *Origin of Species* is as design-like obsessed as is *Natural Theology*.

What about theism? There are certainly going to be issues with Darwinism. The Augustinian atonement theory of salvation – penal substitution – starts with the sin of Adam and Eve, and Darwinism denies that there was a unique first couple. Moreover, at any particular time, sin did not suddenly appear on the scene. Parents and grandparents were just as much sinners. There are ways around this problem, for instance by taking up the older incarnational theory of Irenaeus of Lyons, endorsed by the Eastern Orthodox as well as some western Christians, like the Quakers. Here the death on the cross is not a blood sacrifice – which in any case sounds repellently pagan – but a sign of great love and fellow suffering. Jesus is a friend. No need of an original pair of sinners. However, these are not our direct problems. What about the special status of humans? I see no reason why the Christian believer should not think that God made us in the way sketched in this chapter and leave things at that. We are special, a lot more so than warthogs. There is perhaps the problem of whether a process like evolution through selection could guarantee the arrival of humans. I will leave that problem until later, in the chapter on progress. For now, conclude that perhaps Aubrey Moore really was right.

Humans are of value. This comes not from Darwinism, which is value-drained, but from the Christian perspective on human beings, which is value-loaded. This is why, paradoxically, I see the secular thinker might have more troubles with the Darwinian approach. As Rorty says, we are not that special. "Darwinism requires that we think of what we do and are as continuous with what amoebas, spiders, and squirrels do and are" (Rorty 1998, 295). We are no more than fancy warthogs. Warthogs take cause and effect seriously. Ask them what they are going to do when they hear a lion roar. We may be top dog, but it turns out that we are not really competing at Crufts, where winning really does confer status. At best, we are competing with everyone else's pets in the local park. Relative value rather than absolute value. I will leave this worry now, because it is going to be the focus of the next chapter. But it is a worry. The religious can put in their own values. The secular do not want to do this. They want to find values. And Darwinism, a scientific theory beneath the machine metaphor, simply does not give these.

And so to the creationists, existentialists and others. Apparently, they cannot have everything they want. That total freedom demanded by Sartre – existence precedes essence. We are born into the world and already we have a human nature. I am going to speak to this issue in the Epilogue. For now, like it or not, we are going to think mathematically. Like it or not, we are going to think causally. But truly, whoever thought that we are going to be all that free? You try going through this world without ever eating a meal or going to the lavatory. You try not going in the middle of the night when you are eighty. This all said, creationism in some sense seems plausible. You have the guidelines, the tools, the support. Now go out and use them. Create your own being. Be a painter or a professor or a property manager. Have kids or not. Major in sociology if you must, rather than philosophy. Of course, this presupposes free will, but other than scientists with tin ears for philosophy, whoever thought that science denies freedom?

> For what is meant by liberty, when applied to voluntary actions? We cannot surely mean that actions have so little connexion with motives, inclinations and circumstances, that one does not follow with a certain degree of uniformity from the other, and that one affords no inference by which we can conclude the existence of the other. For these are plain and acknowledged matters of fact. By liberty, then, we can only mean *a power of acting or not acting, according to the determinations of the will*; this is, if we choose to remain at rest, we may; if we choose to move, we also may. Now this hypothetical liberty is universally allowed to belong to every one who is not a prisoner and in chains. Here, then, is no subject of dispute. (Hume 1748)

The great New England theologian Jonathan Edwards (1754) agrees. If, for no good reason, you tear off your clothes in the middle of campus and declare yourself the Archangel Gabriel, you are not free but crazy. "The Will is always determined by the strongest motive." Motives don't come from nowhere. They come from philosophy professors teaching you to aspire to better things for yourself and others. If I did not believe that sincerely, and if I did not believe that this would have some causal effect on my students, I would never have taught for fifty-five years. With joy.

Enough said. Now let's see what the organicists have to say.

5 Organicism and Human Nature

Early Organicists

Organicism lends itself to an evolutionary interpretation. There are exceptions, notably Louis Agassiz. But the essence of the organic approach is that of development, from embryo to full-grown organism. What more natural, then, than to see an analogous process over time with species? Expectedly, we do see much interest by the organicists in embryology, and, as the fossil record was increasingly uncovered and histories – phylogenies – were revealed, there was an immediate move to link the two. Best known is the German post-Darwinian evolutionist Ernst Haeckel, who formulated his "biogenetic" law: "ontogeny recapitulates phylogeny." And what does this add up to? Look at Haeckel's tree of life. (See Figure 13.) We won! Humans are the supreme point of the evolutionary process. We are special, we are unique, because we are right at the top. In an important sense, we are of greater value than any of the other organisms. And this truly is the whole point and purpose of the metaphor. Ask yourself what it is that makes for the Amish survival of old thoughts and ways. Above all it is the sense of community, so fragile in today's industrialized, urban society. If I want a barn built, I hire a contractor. I don't expect our neighbors to turn up and make a day of it, while our wives do the

96

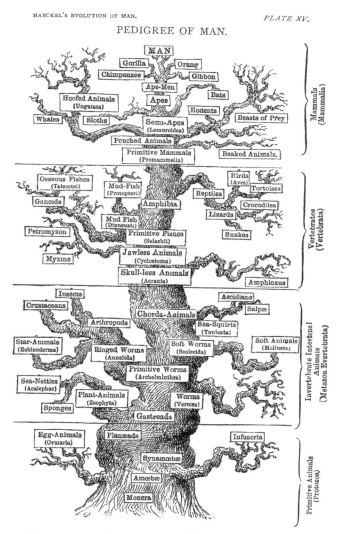

Figure 13 Ernst Haeckel's tree of life from his *The Evolution of Man* (1896).

cooking. For the organicists, exalted human status is the community and barn building.

Organicism/Romanticism may have started in Germany, it spread rapidly. In England, Spencer is the best known. He was not the first. Richard Owen, the anatomist using Platonic themes – all is one – was deeply indebted to Romantic thinking. His idea of the archetype was lifted wholesale (without acknowledgment) from the work of the idealistic morphologist Carl Gustav Carus (Rupke 2009). Owen might have been an evolutionist in the 1830s when Darwin was discovering natural selection. After the *Origin*, Owen became the anti-Christ of the evolutionary world, mainly because it suited Thomas Henry Huxley to pump up the argument; but, in the early years, he and Darwin were good friends, and one much suspects that evolution was discussed in an empathetic way: "every successive animal is branching upwards, different types of organization improving as Owen says" (Darwin 1987, B 19 – this is a personal communication, notebook entry around August 1837). There is even the organism/group analogy.

> All classes of *Acrita* exhibit lowest stages of animal organization, "& are analogous to the earliest conditions of the higher classes during which the changes of the ovum or embryo succeeded each other with the greatest rapidity" – so we find species each class successively present modifications typical of succeeding classes & likewise those much higher in scale. So Owen actually believes in this view!!!" (Darwin 1987, C 48e – notebook entry, quoting Owen, around April 1838).

The post-*Origin* contretemps with Huxley, what Charles Kingsley in *The Water-Babies* called the "great hippopotamus debate," was about the status of humans, with (organicist) Owen claiming that uniquely humans had a hippocampus, and (mechanist) Huxley (1863) arguing that it was possessed by the other primates also. No surprise. On introducing the notion of the archetype in 1849, Owen had written that "Nature" shows advance from "the first embodiment of the Vertebrate idea, under its old Ichthyic vestment, until it became arrayed in the glorious garb of the human form" (86). In the same mode, Herbert Spencer was a fanatical progressionist, writing that "this law of organic progress is the law of all progress. Whether it be in the development of the Earth, in the development of Life upon its surface, in the development of Society, of Government, of Manufactures, of Commerce, of Language, Literature, Science, Art, this same evolution of the simple into the complex, through successive differentiations, holds throughout" (Spencer 1857, 245). He explains that the English language is more complex and hence above all others. We know where his end point lay.

Into the Twentieth Century

The nice thing about this sort of stuff is that you pays your money, and you takes your choice. Spencer put the English at the top. Haeckel put the Germans. Bergson, no doubt, put the French: "not only does consciousness appear as the motive principle of evolution, but also, among conscious beings themselves, man comes to occupy a privileged place. Between him and the animals the difference is no longer one of degree, but of kind" (Bergson 2011, 34). We can put matters more strongly: "in the last analysis, man might be considered the reason for the existence of the entire organization of life on our planet" (35). Whitehead sang the same song. "The whole point of the modern doctrine is the evolution of the complex organisms from antecedent states of less complex organisms"; "the organism is a unit of emergent value"; and so on and so forth (110). No surprise that the whole book leads up to a final chapter on humans and what is needed to keep our progress active.

What of the twentieth-century biologists? The British evolutionist C. H. Waddington (1957) was an ardent Whiteheadian, known scientifically mainly as one who tried to get Lamarckian effects out of Mendelian genetics. His theory of "genetic assimilation" was controversial, not just because of its content, but because he was pushing it around 1950, at a time when Western geneticists were very tense at the power in the Soviet Union of the neo-Lamarckian Trofim Lysenko. Actually, it was really more bark than bite. Waddington subjected fruit flies to extreme, shocking conditions, causing anomalies in growth, and then he selected from the anomalies. Before long, he got the anomalies without need of environmental disturbance, the shocks. Truly it is all pretty Mendelian – you are selecting from those with existing predispositions and gathering them together, so it is little surprise that the combined effects should start showing. In any case, this is not true Lamarckism, for the effects of the shocks – wing distortions – are not of adaptive value, as they are in true Lamarckism, the stronger arm of the blacksmith for instance.

This work was going on against a background of nigh-fanatical organicism. "The capacity to remain relatively independent of the environment, to incorporate into the life-system more complex functions of environmental variables, and ultimately to control the environment, of course reach a much higher point in man than in any pre-human species" (Waddington 1960, 137). Adding: "it would be impossible to conceal that the evolution of the human race has carried forward in many respects sequences of change which we can see running through the sub-human world." No surprise either that this is reflected in cultural evolution: "just as in animal evolution as a whole we can see some direction which justifies us as speaking of certain groups as lower and of others as higher, so when we look at the whole of human development we can see a general pattern of change, from small groups of nomads or scattered communities of food gatherers, to the complex and elaborate civilizations typified by such individuals as, let us say, Confucius, Plato, Newton, and Leonardo" (Waddington 1962, 110–11).

Spandrels

The Harvard paleontologist Stephen Jay Gould never liked to do anything that did not create a stir. With his "Spandrels of San Marco" (1979), co-authored with fellow Harvard evolutionist Richard Lewontin, he certainly succeeded. Edward O. Wilson's *Sociobiology: The New Synthesis* (1975) caused a huge controversy, with Gould and Lewontin very prominent public critics of their Harvard co-departmental colleague. They were highly negative about Wilson's bringing human nature within the Darwinian paradigm. Craftily, they went at matters sideways by criticizing Darwinism as such. At least, arguing that undue focus on natural selection detracted from important organic characteristics demanding non-Darwinian explanations.

> An adaptationist programme has dominated evolutionary thought in England and the United States during the past 40 years. It is based on faith in the power of natural selection as an optimizing agent. It proceeds by breaking an organism into unitary "traits" and proposing an adaptive story for each considered separately. Trade-offs among competing selective demands exert the only brake upon perfection; non-optimality is thereby rendered as a result of adaptation as well. We criticize this approach and attempt to reassert a competing notion (long popular in continental Europe) that organisms must be analyzed as integrated wholes, with Baupläne so constrained by phyletic heritage, pathways of development and general architecture that the constraints themselves become more interesting and more important in delimiting pathways of change than the selective force that may mediate change when it occurs. (Gould and Lewontin 1979, 581)

Anti-adaptation, anti-natural selection, anti-reduction, pro-Baupläne (akin to Owen's archetypes), pro-holism, pro-development (embryology), pro-European biology. Gould was a master rhetorician and so, while he was busily demolishing all that Darwin held dear, he tells us that he is working in the spirit of Darwin who would, undoubtedly, be pleased at what he is doing!

> Since Darwin has attained sainthood (if not divinity) among evolutionary biologists, and since all sides invoke God's allegiance, Darwin has often been depicted as a

radical selectionist at heart who invoked other
mechanisms only in retreat, and only as a result of his
age's own lamented ignorance about the mechanisms of
heredity. This view is false. (589)

Tucked away at the end of this part of the discussion is an
interesting qualification. "We do not now regard all of
Darwin's subsidiary mechanisms as significant or even
valid"! Nevertheless, "we should cherish his consistent atti-
tude of pluralism in attempting to explain Nature's complexity." This is known in the trade as "damning with faint
praise." Overall though, back to morphology, back to archetypes, back to non-adaptive evolutionary processes!

> Under the adaptationist programme, the great historic
> themes of developmental morphology and Bauplan were
> largely abandoned; for if selection can break any
> correlation and optimize parts separately, then an
> organism's integration counts for little. Too often, the
> adaptationist programme gave us an evolutionary
> biology of parts and genes, but not of organisms. It
> assumed that all transitions could occur step by step and
> underrated the importance of integrated developmental
> blocks and pervasive constraints of history and
> architecture. A pluralistic view could put organisms, with
> all their recalcitrant, yet intelligible, complexity, back into
> evolutionary theory. (597)

Although not mentioned in this paper, it was just at this time
that Gould was pushing his alternative to Darwinism, non-
adaptive jumps from one form to another – "punctuated
equilibrium" (Gould and Eldredge 1977). The Spencerian
resonances of both the language and the claims hardly need

mentioning. Also not mentioned, at this time Gould was committed to a thoroughly progressionist view of evolution, with humankind the winners at the top. His very long *Ontogeny and Phylogeny* (1977) was an invitation to resurrect and revalue Haeckel's biogenetic law. He made the case for progress in terms of the distinction between r-selection – where organisms have lots of offspring but give them little care (herrings) – and K-selection – where organisms have few offspring and give them much care (primates).

> I have been trying to deemphasize the traditional arguments of morphology while asserting the importance of life-history strategies. In particular, I have linked accelerated development to r-selected regimes and I have identified retarded development as a common trait of K-strategists. . . . I have also tried to link K selection to what we generally regard as "progressive" in evolution, while suggesting that r selection generally serves as a brake upon such evolutionary change. I regard human evolution as strong confirmation of these views. (Gould 1977, 504)

Darwin, incidentally, great revolutionary but no rebel, made the case in terms of the Irish (r-selected) and the Scots (K-selected).

Sewall Wright

These organicists show a common pattern of playing fast and loose with natural selection. They are for it and then suddenly they are not. Perhaps the best case of all is the hugely important and influential American population geneticist Sewall Wright. He was always a bit half-hearted about natural selection. He was not at all sure it could produce the needed change for evolution. Most revealingly, at a National Research Council conference in 1955 to discuss biological concepts, participants were asked to say which concepts they thought most significant. Wright tagged "organization," "replication," "variation," "evolution" (the fact not the cause), and "hierarchy." He (as did most of the participants) ignored "natural selection," chosen by Mayr and Simpson, two other evolutionists at the conference.

Wright's "shifting balance" theory of evolution supposed that populations (species) are in equilibrium. Something untoward happens to fragment them into small groups where random effects can outweigh selection (drift). It is here that new innovations come into being, by chance as it were. The small groups then join up and there is a kind of intra-specific group selection, as more successful groups (thanks to their randomly obtained innovations) take over and become the species norm, back to equilibrium. If this doesn't start to ring a bell, consider the comment Wright makes at the end of the paper introducing this theory. "The present discussion has dealt with the problem of evolution as one depending wholly on mechanism and chance. In recent years, there has been some tendency to revert to more or less mystical conceptions revolving about such phrases as 'emergent evolution' and 'creative evolution'" (Wright 1931, 155). He rushes to say that he thinks such ideas have no place in science, but he "must confess to a certain sympathy with such viewpoints philosophically." He certainly did, for his theory is Herbert Spencer in mathematical form. No surprise at all, because Wright was a graduate student at Harvard in the second decade of the twentieth century, when the biology faculty – including his own father – were ardent Spencerians to a man.

Philosophers for Organicism

What about the philosophers? The French philosopher Gilles Deleuze was an enthusiastic organicist, somewhat expectedly, given the influence on him of Henri Bergson (Bennett and Posteraro 2019). The Anglophones? We have seen that they are enthusiastically committed to the special significance of humans. Very revealingly, John Dupré puts genetic drift up at the same level as natural selection. "Conceptually, selection and drift are quite different processes, but in practice they can be extremely difficult to separate. Once we see that the trajectory of a population through time is one in which adaptedness is always maintained, the conditions for existence are continuously met, it is very difficult to distinguish among the causes of this maintenance" (Dupré 2010). In any case, at best, natural selection does little. "Where does adaptive change come from? A trivial but sometimes obfuscated point is that it never comes from natural selection." Continuing: "Selection cannot occur unless some other process provides alternatives to select from. It follows that any thesis about the power of natural selection to generate change implicitly presupposes a thesis about a process or processes that generate selectable change" (Dupré 2017). In other words, the hard work is being done by mutation, not the Darwinian contribution of selection.

Dupré makes few bones about the fact that he has little time for the mechanical world picture. "There are powerful reasons for thinking that emancipation from the mechanistic paradigm is a precondition for true insight into the nature of biological processes" (Dupré 2012a, 83). No prizes for guessing where this is going to take us: "there are limits as to how far conventional mechanistic explanations can take us in understanding the dynamic stability of processes at this hierarchy of different levels. Such understanding will require models that incorporate both the capacities required by mechanistic or quasi-mechanistic constituents, and the constraints and causal influences provided by properties of the wider systems of which these constituents are parts" (203). All of this is at one with the general philosophical distaste for reductionist thinking. "Traditional reductionist views of science, with their focus on 'bottom-up' mechanisms, do not suffice in the quest to understand top-down and circular causality and a world of nested processes" (Dupré 2012b).

Fodor tells us he hopes for new paradigms. Really, his main hope seems a return to organicism. He positively wallows in embryology and its significance for evolutionary change. Of the correct evolutionary picture, we learn: "The slogan is the evolution of ontogenies. In other words, the whole process of development, from the fertilized egg to the adult, modifies the phenotypic effects of genotypic changes, and thus 'filters' the genotypic options that ecological variables ever have a chance to select from" (Fodor and Piattelli-Palmarini 2010, 27). And that of course is precisely what the Romantics claim. Look at the evolution of the individual and you have the answer to the evolution of the group.

Completing the trinity we have Thomas Nagel (2012). Unlike Dupré, who with his enthusiasm for genetic drift downgrades the design-like nature of the organic world, Nagel finds the design-like nature of the organic world overwhelming. So much so that the Darwinian attempt to explain this phenomenon convinces him that we may need to step right outside the mechanistic paradigm. He speculates that possibly "there are natural teleological laws governing the development of organization over time, in addition to laws of the familiar kind governing the behavior of the elements." He allows that: "This is a throwback to the Aristotelian conception of nature, banished from the scene at the birth of modern science. But I have been persuaded that the idea of teleological laws is coherent, and quite different from the intentions of a purposive being who produces the means to his ends by choice. In spite of the exclusion of teleology from contemporary science, it certainly shouldn't be ruled out a priori" (22).

Two Root Metaphors

Enough said. I want to stress that, although I am not an organicist, I am not presenting these arguments to criticize them. Anything but. The whole point of the exercise is to show how serious thinkers reject mechanism and the jewel in its crown, Darwinian evolutionary biology. Of course, the Darwinians don't much care for the arguments. One of the leading English evolutionary biologists of the 1950s, Arthur J. Cain, a member of Ford's school of ecological genetics, in conversation with me, punning on the English slang term "codswallop," meaning total nonsense, always referred to Waddington's work as "wadscollop." John Maynard Smith, also a leading English evolutionist, was scathing about Gould.

> Gould occupies a rather curious position, particularly on his side of the Atlantic. Because of the excellence of his essays, he has come to be seen by non-biologists as the preeminent evolutionary theorist. In contrast, the evolutionary biologists with whom I have discussed his work tend to see him as a man whose ideas are so confused as to be hardly worth bothering with, but as one who should not be publicly criticized because he is at least on our side against the creationists. All this would not matter, were it not that he is giving non-biologists a largely false picture of the state of evolutionary theory. (Maynard Smith 1995, 46)

In the same mode, Dupré's speculations get short shrift in today's biology departments. "We do not need a new philosophical framework for evolution, much as Dupré wants one. Traditional reductionist views are still valid and yielding valid insights (what is microRNA other than a 'bottom-up' phenomenon that regulates genes?)." Jerry Coyne adds: "As an evolutionary biologist – which Dupré is not – I think I'd know if my field was in crisis. Yet I haven't heard any recent lamentations from my colleagues" (Coyne 2012). Take genetic drift. It is thought very important at the molecular level, beneath the force of natural selection. It is this fact that enables biologists, very accurately, to date past events – like when did humans and chimpanzees split? As something brought on by interactions at the physical (phenotypic) level it is virtually discounted (Coyne, Barton, and Turelli 1997).

Famously, in the 1930s, Theodosius Dobzhansky accepted drift as the explanation for the genetic differences he found out West in isolated populations of fruit flies. Then he discovered that, because the phenotypes were differentially selected over the seasons of the year, the genetic differences varied in tandem. He swung at once to a Darwinian explanation, from which he never varied. That was why, as one now fully Darwinian, it was so important to Dobzhansky to establish variation as the norm in natural populations. The irony will not be lost on the reader that it was Jerry Coyne's doctoral supervisor, Richard Lewontin, co-author of the spandrels paper, who, through ultra-reductionist techniques, showed that there is such variation in populations and hence the potential of natural selection is made hugely more plausible (Lewontin 1974).

As I say though, at one level, I am not so much interested in who is right and who is wrong. I am more concerned to suggest that this total breakdown in understanding and sympathy suggests something more than mere factual or theoretical difference. We have an abyss. It is a matter of different root metaphors. We saw that Thomas Kuhn identified paradigms with metaphors (Kuhn 1993). We have two competing conceptual frameworks. Note, not evolutionary versus non-evolutionary. Organicists enthuse about development as much as mechanists. It is a question of Darwinism-external-forces-shaping-organisms/mechanism versus Romanticism-internal-drive-shaping-organisms/organicism. Endorsing one rather than the other is a bit like being a Catholic or a Protestant, a Republican or a Democrat. You can bring all sorts of arguments to bear – I doubt Kuhn's claim we have ontologically different worlds – but ultimately it is a matter of faith commitment. No Protestant is going to talk a Catholic out of transubstantiation. No Democrat is going to talk a Republican out of the conviction that the defense of free enterprise must always come first and foremost. Kuhn suggested that paradigms are temporal – one replaces another – but there is no reason why they should not persist. Of course, Kuhn was thinking mainly of specific theories rather than all-encompassing world pictures, but if the shoe fits, wear it.

Arguing about right and wrong is going to get you nowhere. However, we are not yet at an unpassable barrier. As a pragmatist, one has a further step to take. One judges which metaphor/paradigm does a better job, in the sense of tackling problems and coming up with answers. Here at once we find the reason why someone like me opts for mechanism. You cannot do as much good science under the organic metaphor as under the machine metaphor. This applies to the life sciences as much as to the physical sciences. While the Whiteheadians were into late-night, philosophical bull sessions, meaningfully talking about organization, James Watson and Francis Crick were at work in Cambridge. Hard to imagine two less-ethereal human beings. When their minds were not on science, they were on girls or pints in the pub. Firmly committed to the machine metaphor, they uncovered the double helix, arguably the most fertile scientific breakthrough of the twentieth century. In evolutionary studies also. The science of the behavior of organisms from a Darwinian standpoint, sociobiology, raced ahead precisely because evolutionists devised a number of individual-selection models, notably kin selection and reciprocal altruism, hinted at by Darwin but far from fully developed, that simply transformed the field. What was hitherto puzzling now became clear. And no amount of going on about organization or how selection can work on species as opposed to individuals – group selection over individual selection – changed that. Holism threw no light on the caste systems of the hymenoptera. Reductionism – take the machine apart and look at the pieces and see what they do – did. William Hamilton (1964a, 1964b), with his brilliant arguments that, in helping at the nest, non-reproductive workers were ensuring that copies of their own genes get into the next generation, showed in one stroke that "sacrifice for the good of others" arguments are neither needed nor helpful.

As a codicil one might add that, if Darwinian selection is downgraded as the main cause of evolution, the organicists have got to provide an alternative. The tradition, illustrated by Fodor, is to seek it in embryological development. Somehow, the development of the organism supposedly yields the required change, without direct aid from outside. Evolutionary development ("evo-devo") is today the promising field of study and "epigenetics" the key word. Enthusiasts are open in their enthusiasm for the significance of ontogeny (individual development) and their disdain for natural selection.

> The homologies of process within morphogenetic fields provide some of the best evidence for evolution – just as skeletal and organ homologies did earlier. Thus, the

ORGANICISM AND HUMAN NATURE

evidence for evolution is better than ever. The role of
natural selection in evolution, however, is seen to play less
an important role. It is merely a filter for unsuccessful
morphologies generated by development. Population
genetics is destined to change if it is not to become as
irrelevant to evolution as Newtonian mechanics is to
contemporary physics. (Gilbert, Opitz, and Raff 1996, 368)

Darwinian mechanists, needless to say, are no less enthusiastic about evo-devo, seeing it as an ultra-reductionistic,
molecular approach to uncovering the mysteries of the way
the genes – the genotype – program development up to full-
blown organisms – the phenotype (Futuyma 2017). Just as
the work of Hamilton and other sociobiologists plugged a
gap about social behavior as treated by Darwinian evolutionary theory, so those working on development plug the gap –
the "black box" – in Darwinian evolutionary theory taking
us from selection-driven gene ratios to the evolution of
populations of full-grown organisms. Far from being
rendered irrelevant, natural selection is seen as being as
important as ever. Moreover, Darwinians point out that,
even if evo-devo proves a vital element in showing the
superiority of the organicist approach, there is still a long
journey ahead showing exactly how the details support the
desired conclusions. If you are not going to explain final
cause mechanistically, how then are you going to explain it?

Humans as Special

We swing back to the three-fold way of Chapter 1. The
organicists are not pragmatists. My way of settling things

cuts no ice. It is the metaphysics that counts. Humans are special and their metaphor makes this central. Mechanism does not. You should not sell your birthright for a mess of nineteenth-century, New England, philosophical potage. What then of the status of humans as promoted by organicism? The religious can be mechanists. For the Christian – less for the Buddhist I suppose – there are tensions with evolutionary thought, about the literal existence of Adam and Eve for instance. I would suppose, however, that these tensions are no less for the organicist. Organicism is evolutionary, and that includes human beings. In fact, the whole attraction of organicism is that it does include human beings and they emerge naturally from the lower animals. There is going to be no unique pair bringing sin into the world for the first time. This said, many believers do find the organicist position more congenial. Process theologians (followers of Whitehead) especially: "it is now clear that mechanism by itself cannot be an adequate way to understand organisms" (Birch and Cobb 1981, 72). God did not leave things to chance, hoping that humans would emerge from the evolutionary process. In some sense – in the sense that, once the ovum is fertilized, you are on the way to the adult organism – it was all preplanned. Humans were bound to emerge and be the peak of the whole process. We are products of the system. "*Homo sapiens* is part of nature." But we are different. "Human beings play a very distinctive role, and they have separated themselves a long way from all the other products of the evolutionary process" (97). Just the job for Christian doctrine. Our powers of reason and thought come with the territory.

114

They are natural in some sense. This is not chance. We had to think in the ways that we do.

How does the secular thinker regard organicism? I presume, as is shown by people like Dupré and Nagel, the organic model is precisely what is needed. Humans are of value, the greatest value. Mechanism – Darwinism – is drained of value, at least in any absolute sense. You must put value into the picture. However, value is part of the organicist package. The world is of value, and that value increases as we get closer to human beings. What about consciousness and that sort of thing? Nagel criticizes Darwinism for not being able to get a handle on it. Since Darwinian evolution cannot solve the mind–body problem, this is a devastating consequence for natural selection. Of thinking animals – Nagel does not deny that there are non-human, thinking animals, he just takes it for granted that they are not our equivalent – he argues that: "An account of their biological evolution must explain the appearance of conscious organisms as such" (Nagel 2012, 45). Continuing: "Selection for reproductive fitness may have resulted in the appearance of organisms that are in fact conscious, and that have the observable variety of different kinds of consciousness, but there is no physical explanation of why this is so – nor any other kind of explanation that we know of" (46).

Perhaps surprisingly, I agree with Nagel entirely! But I would argue that the problem – if such it be – lies not with Darwinism as such but with the machine model generally. Machines cannot think. If you go with mechanism, in that sense you are ruling mind out of the picture entirely. Leibniz pointed this out in the *Monadology*.

> One is obliged to admit that perception and what depends upon it is inexplicable on mechanical principles, that is, by figures and motions. In imagining that there is a machine whose construction would enable it to think, to sense, and to have perception, one could conceive it enlarged while retaining the same proportions, so that one could enter into it, just like into a windmill. Supposing this, one should, when visiting within it, find only parts pushing one another, and never anything by which to explain a perception. Thus it is in the simple substance, and not in the composite or in the machine, that one must look for perception. (Leibniz 1714, 215)

Since the Darwinian approach to philosophy presupposes the mechanistic perspective, the mind–body problem is insoluble. This is not necessarily the knife that strikes to the heart of Darwinism, leaving it stone dead. That science cannot answer every question comes with the reliance on metaphor. As Kuhn stressed, paradigms/metaphors are like the blinkers on horses. They work by pushing certain topics out of the realm of discourse or understanding. We have seen how the machine metaphor pushes out value questions. It also has nothing to say about what Heidegger called the fundamental question of metaphysics. Why is there something rather than nothing? When thinking about machines, you don't spend all of your time worrying about where the materials came from. First take your hare. Same with the mind–body problem. Machines don't think.

This doesn't mean that you cannot say anything or that the Darwinian approach cannot contribute. Anatomically, you can say a huge amount about what parts of the brain support what types of thinking and abilities. Also, you can say what kind of philosophy of mind fits most readily or easily with Darwinism. Against Nagel, the title of whose book tips Darwinism into the materialist camp, many Darwinians favor some sort of monism, body and mind are one. Indeed, they generalize, thinking mind pervades the universe – "pan-psychic monism." Phenomena like quantum entanglement, where information can be transmitted instantaneously across the universe, suggest that perhaps mind is more pervasive than we think. The late-nineteenth-century philosopher-mathematician William Kingdom Clifford was ahead of us.

> We cannot suppose that so enormous a jump from one creature to another should have occurred at any point in the process of evolution as the introduction of a fact entirely different and absolutely separate from the physical fact. It is impossible for anybody to point out the particular place in the line of descent where that event can be supposed to have taken place. The only thing that we can come to, if we accept the doctrine of evolution at all, is that even in the very lowest organism, even in the Amoeba which swims about in our own blood, there is something or other, inconceivably simple to us, which is of the same nature with our own consciousness, although not of the same complexity. (Clifford [1874] 1901, 38–39)

I don't think this solves the mind–body problem. It does set the parameters. You are not now making mind separate from the machine, Leibniz's object of attack, but making mind part of the material of the machine. I don't see that this necessarily introduces absolute values. Mind is universal.

Incidentally, organicism, being also based on a metaphor, should not feel itself too superior to mechanism with respect to questions unanswerable. You can go to your local aggie college as often as you like for superior forms of hybrid corn, but ultimately it all starts with the Great Seed Store in the Sky. Likewise with the mind–body problem. Interestingly, but not so very surprisingly, panpsychic monism is no less attractive to the organicist. Whitehead was ever an enthusiast and so also was Sewall Wright. Indeed, Nagel himself seems not entirely negative. There are still problems to be solved, notably the combination problem. How do you get the minds of individual molecules to come together to make one human functioning mind – Charles Darwin or Herbert Spencer? What does seem to be true is that organicism is a piece of candy for the secularist who wants to make – or rather find – humans special. I am not surprised that many secularists realize this.

Finally, the creationist option. Existence precedes essence. We put the meaning into life. I suppose you could be an organicist, although I am not quite sure how comfortably it sits with an existentialist like Sartre. He wants to create value, rather than have it there given to us, as is the case with organicism. In a way it is hard to give a definitive answer here. I doubt anyone before has ever asked if an organicist can be an existentialist! But that is my take on things. I will pick up again on this question in the Epilogue.

Progress

There is one final point which needs raising. Am I not selling mechanism-Darwinism short? A key factor here is the organicists' conviction that they see progress, a value-conferring and -increasing phenomenon, leading up to the arrival of human beings. Is this so entirely outside the scope of Darwinism? Richard Dawkins seems to think you can get progress and, if anyone is a hardline Darwinian, it is he. Remember: "Directionalist common sense surely wins on the very long time scale: once there was only blue–green slime and now there are sharp-eyed metazoa" (Dawkins and Krebs 1979, 508). The same is true of another prominent Darwinian, the author of the *Origin of Species*! This is from the third edition of the *Origin* (1861).

> If we look at the differentiation and specialisation of the several organs of each being when adult (and this will include the advancement of the brain for intellectual purposes) as the best standard of highness of organisation, natural selection clearly leads towards highness; for all physiologists admit that the specialisation of organs, inasmuch as they perform in this state their functions better, is an advantage to each being; and hence the accumulation of variations tending towards specialisation is within the scope of natural selection. (134)

And remember that conclusion to the *Origin,* with its talk of "grandeur in this view of life," where "from so simple a beginning endless forms most beautiful and most wonderful have been, and are being, evolved" (490).

Most interesting is the evolutionist Theodosius Dobzhansky, introduced already as one of the great shapers of modern evolutionary thinking. He was a Christian, a member of the Russian Orthodox Church. He was also, amazingly, the president of the American branch of the

Teilhard de Chardin Society. Teilhard was the French Jesuit priest, a brilliant paleontologist, who used the Bergsonian philosophy to argue that evolution is progressive, up through humankind to the Omega Point, which he identified with Jesus Christ (Teilhard de Chardin 1955). Dobzhansky was openly a progressionist seeing a process up to humans. This is from a letter to the historian of science John Greene, also a Christian, but not a Teilhardian.

> Certain evolutionary processes are "creative" because they bring about (a) something new (b) having an internal coherence since it maintains or advances life, and (c) may end in either success or failure. One of the notable successes, let us say the greatest success, was the origin of man. (Greene and Ruse 1996, 457)

My sense is Dobzhansky was not entirely clear how this happens. He was a Darwinian and wanted to stay in the machine metaphor world. Yet, although he did not much like Whitehead's philosophy, he was drawn to a Whiteheadian position.

> I see no escape from thinking that God acts not in fits of miraculous interventions, but in all significant and insignificant, spectacular and humdrum events. Panentheism, you may say? I do not think so, but if so then there is much truth in panentheism. (463)

Panentheism, which goes back to Schelling and was a significant part of Whitehead's philosophy, differs from pantheism, which identifies God with the world, in seeing God separate from but pervasive through His creation. Process theology pushes this to extremes that must have St.

Augustine revolving rapidly in his grave. God is so much part of the process that as he affects the world, so the world affects Him, making him part of the evolutionary process. "God not only acts on others, but also takes account of others in the divine self-constitution" (Birch and Cobb 1981, 196). Hence: "God perfects the world and the world perfects God" (197).

I am not sure – I doubt Dobzhansky was sure – about whether this all holds together. One suspects he recoiled in horror at the just-quoted consequences of process theology. Leave it as mechanism leading to progress with God behind it all. Perhaps that is the option that must be taken by any would-be Christian Darwinian. Let us, however, shelve that question, until we have asked the prior question. Can Darwinism yield biological progress up to humankind? This is the topic of the next chapter.

6   The Problem of Progress

The Question of Progress

The idea of progress at the cultural level is that of humans making things better – education, health, material comfort, safety, and so forth (Bury 1920). It is important to stress that it is humans making things better. The very essence of progress is that we do it ourselves. Christians traditionally have had a rival ideology, Providence. This is the idea, in its usual Augustinian form, that everything is due to God – His grace – and everything good comes through the Blood of the Lamb. We humans are helpless, mired in original sin, and save for the sacrifice on the cross we are doomed to ever-lasting misery. Whereas progress stressed it is all up to us, Providence insisted that we unaided could do nothing.

> Amazing grace, How sweet the sound
> That saved a wretch like me.
> I once was lost, but now I am found,
> Was blind, but now I see.

Things really started to change by the eighteenth century, the Age of the Enlightenment. Thanks to science, to increasing prosperity, to ever-more-secular philosophical analysis, increasingly there was the conviction that we don't need God so much. We can do things ourselves. Progress started to edge out Providence. And it is here that evolution starts to come into the story (Ruse 1996). As the Romantics asked, could it be that as individual organisms develop, so whole groups might develop analogously? What would this mean? Most obviously, from the simple to the complex, from the undesirable to the desired, from the blob to the human? Even before Romanticism started to take form, several started to speculate in this way. This is Denis Diderot, the French philosopher and creator of the huge compendium of knowledge, the *Encyclopédie*, writing in the middle of the eighteenth century: "Just as in the animal and vegetable kingdoms, an individual begins, so to speak, grows, subsists, decays and passes away, could it not be the same with the whole species?" (Diderot 1943, 48) Here he uses the organic analogy, but, in this pre-Romantic era, it was truly the cultural progress/biological progress that was fundamental. Note that, since this analogy is something that comes out of science and technology, for Diderot and like thinkers, biological progress is located more under the machine metaphor than the organic metaphor.

Keeping it in the family, let us turn to Erasmus Darwin, late-eighteenth-century physician, grandfather of Charles Darwin, who was an enthusiast for taking Progress in the cultural world – from now on I shall follow convention and speak of the cultural notion as Progress, with the P capitalized – and applying it analogically to the organic world, progress – the biological notion, without a capital p. Breaking into verse.

> Organic Life beneath the shoreless waves
> Was born and nurs'd in Ocean's pearly caves;
> First forms minute, unseen by spheric glass,

Move on the mud, or pierce the watery mass;
These, as successive generations bloom,
New powers acquire, and larger limbs assume;
Whence countless groups of vegetation spring,
And breathing realms of fin, and feet, and wing.
Thus the tall Oak, the giant of the wood,
Which bears Britannia's thunders on the flood;
The Whale, unmeasured monster of the main,
The lordly Lion, monarch of the plain,
The Eagle soaring in the realms of air,
Whose eye undazzled drinks the solar glare,

Imperious man, who rules the bestial crowd,
Of language, reason, and reflection proud,
With brow erect who scorns this earthy sod,
And styles himself the image of his God;
Arose from rudiments of form and sense,
An embryon point, or microscopic ens!

(Darwin 1803, 1, 11, 295–314)

Biological progress! And to underline the point, explicitly Darwin tied biological progress in with cultural Progress. The idea of organic progressive evolution "is analogous to the improving excellence observable in every part of the creation; such as the progressive increase of the wisdom and happiness of its inhabitants" (Darwin 1794–96, II, 247–48). Again, a mechanist's view of progress, not that of the organicist.

Progress, cultural, continued to underpin progress, biological. As the nineteenth century got under way, this melded in with the organic analogy, and it was the latter that became dominant. Herbert Spencer saw Progress everywhere, complementing his beliefs in progress everywhere.

Moving rapidly forward, just before the middle of the twentieth century, Julian Huxley waxed strong on the subject. He was always a fanatical Progressionist and for the life of him could not keep progress out of his biology writings. In his major overview of the field – *Evolution: The New Synthesis* (1942) – there was progress, front, back, and throughout. A no-nonsense, brutally strong progress. "One somewhat curious fact emerges from a survey of biological progress as culminating for the evolutionary moment in the dominance of *Homo sapiens*. It could apparently have pursued no other general course than that which it has historically followed" (Huxley 1942, 569). Not that we are the end point.

> The Ant herself cannot philosophize –
>> While Man does that, and sees, and keeps a wife,
> And flies, and talks, and is extremely wise . . .
> Yet our Philosophy to later Life
>> Will seem but crudeness of the planet's youth,
>> Our Wisdom but a parasite of Truth.

This is from a sonnet, "Man the Philosophizer." Huxley's mother was the niece of Matthew Arnold. Not all change is Progress.

To bring the story up to the present, we have Edward O. Wilson and his fervent belief in biological progress. "The overall average across the history of life has moved from the simple and few to the more complex and numerous." Teasing apart mechanist- versus organicist-inspired notions of progress is a job to be taken up. We can say that modern writers stress the interconnections of culture and biology, where cultural Progress is not just an analogy for biological progress, but where culture and biology are intertwined in one overall P/progressive picture. We learn from evolutionary biologist Daniel E. Lieberman (2013) that four or so million years ago there were climate changes, with less forest and more savannah. There was therefore ecological space to branch out – which our ancestors did, unlike those of the other great apes. Whether we moved out from choice or because of necessity given that others were more successful in the forests is a good question.

We know that a lot of characteristically human features, particularly bipedalism and the freeing up of the forelimbs and hands, came because of these moves. Bipedalism is not that good for fast running – literally, my four-legged cairn terriers run circles around me – but being up on two legs does make possible ongoing running for distances without breaks. You may not be able to outrun your prey, but you can keep going until literally they drop dead from exhaustion. Helping is the fact that being upright cuts down on the amount of the body exposed to solar rays. Unlike the prey, you are not overheating from the sun. Whether as cause or effect, or a combination of both, the first proto-humans were hunter-gatherers – out after game (mainly male) or picking fruits and berries and roots (a big job for females). They were notable for their cooperation, working together and sharing. Other great apes are not much into this. Most probably the growth in brain size was part and parcel of the causal process – bigger brains mean more efficient hunter-gathering. More efficient hunter-gathering means ever more success biologically.

All of this at first was confined entirely to Africa, and then 2 mya bands started to venture forth, finally covering the whole globe. Modern humans appeared about 200,000 years ago and moved out of Africa into the Middle East about 80,000 years ago. Things start to pick up about 50,000 years ago, which is when modern *Homo sapiens* started to spread around the globe. There was development of language, more sophisticated tools, and so forth. With these moves and development, we start to see the evolution of distinguishing features, notably skin color. Near the equator you need dark skin to counter the ill effects of ultraviolet radiation; in more temperate climes, you need light skin to produce vitamin D. The big breakthrough to agriculture came about 13,000 years ago. An Ice Age struck, the hunter-gatherer lifestyle was no longer that sustainable, populations that had started doing a bit of cultivation and the like proved to be at a great selective advantage, and the rest as they say is history.

Biologically, agriculture is a smash hit. On the one hand, there were animals and plants available to be domesticated and humans took advantage of this. Cattle, sheep, pigs, chickens, rice, wheat, potatoes. Often the relationship became symbiotic. Teosinte had just a few kernels that could when ripe detach themselves. As it was bred more and more extensively and became what we (in Britain) call maize or (in America) corn, it required more and more human help to detach the kernels from the cob. On the other hand, we learnt to make and fashion helpful tools and artifacts. Pottery comes into existence. Grinding stones start to appear. What is the result? It is simple – "farmers pump out babies much faster than hunter-gatherers" (Lieberman 2013, 188). We tend to think of large families as drains on the resources. For farmers, it is the opposite. "After a few years of care, a farmer's children can work in the fields and in the home, helping to take care of crops, herd animals, mind younger children, and process food. In fact, a large part of the success of farming is that farmers breed their own labor force more effectively than hunter-gatherers which pumps energy back into the system, driving up fertility rates."

In 10,000 years, the number of humans increased 100-fold – from about 5 or 6 million when agriculture got under way to about 600 million at the time of Jesus. And obviously this was just a start. We keep going, adding almost as many again to the beginning of the nineteenth century. (Then things really took off, with around seven or eight billion today.) There is really no need to labor the point. Humans have been a huge success. And if you keep thinking at the purely biological level, as Al Jolson told us truthfully, you ain't seen nothing yet. The human population is growing at 83 million a year. Incidentally, don't think that truly all of this P/progress is really just at the cultural level. There is ongoing feedback. The move to agriculture meant the availability of milk products. Normally, adult humans are lactose-intolerant and cannot digest such products. The right genes came along and made all possible. As needed. The Irish are lactose-tolerant; in Asia, where agriculture did not yield these products, people are still lactose-intolerant. Progress leading to progress, leading back to Progress.

All in all, there is no surprise that, in the general domain also, biological progress is alive and well. Not just in the laboratory of Edward O. Wilson, but in the popular mind, in the way that evolution is presented in print or on

Figure 14 The standard picture of human evolution.

the radio or television. The course of biological history is almost inevitably represented as one leading directly to *Homo sapiens* (Figure 14).

The Naysayers

What about Darwinian evolutionary theory and the warthoggian issue? Does Darwinian evolutionary theory tell you that humans are superior to warthogs? Stephen Jay Gould was rhetorically flamboyant on the topic. No surprise. By the 1980s, biological progress was "a noxious, culturally embedded, untestable, nonoperational, intractable idea that must be replaced if we wish to understand the patterns of history" (Gould 1988, 319). Big surprise. Gould saw no inevitability to the emergence of humans. Making a facetious reference to the asteroid that hit the Earth 65 million years ago, thereby wiping out the dinosaurs and making possible the proliferation of the hitherto rat-like, nocturnal mammals, he wrote: "Since dinosaurs were not moving toward markedly larger brains, and since such a prospect may lie outside the capabilities of reptilian design (Jerison

1973; Hopson 1977), we must assume that consciousness would not have evolved on our planet if a cosmic catastrophe had not claimed the dinosaurs as victims. In an entirely literal sense, we owe our existence, as large and reasoning mammals, to our lucky stars" (Gould 1988, 318).

F. Scott Fitzgerald said: "The test of a first-rate intelligence is the ability to hold two opposed ideas in mind at the same time and still retain the ability to function." Gould qualifies. Truly, though, Gould was not really that Hegelian. The sociobiology controversy convinced Gould that thoughts of biological progress lead to racism. Humans are not only top, but certain groups of humans are even toppier. The American immigration laws of the 1920s, essentially ending immigration by European Jews, proved the point. As a child of such Jews, Gould wanted nothing of it (Gould 1981). So, he wrote often and passionately against biological progress. And, to be fair, his thinking against biological progress was based on the very essence of Darwinism. Gould as non-progressionist mechanist, as opposed to Gould as progressionist organicist. The building blocks of change, variations or, in terms of genetics, mutations, are random in the sense of not appearing as needed. Pollution changes the background from white to black and so you need a new color for your camouflage. Black instead of white. You might as likely get bright red or psychedelic green. And, what might be available and what might work is a veritable crap shoot. Not much direction there.

Then there is natural selection itself. It is relativistic. Being large is not inherently good in itself. Elephants may be free from predators, but they need a lot of roughage as feed.

The hobbit on the island of Flores makes the point. Food was scarce so there was a premium on being small. If the island were overflowing with fruit and vegetables, but its denizens threatened by indigenous predators, the premium would be on being large. There is no inevitable change in one predetermined desirable way. Darwin's theory seems to have built-in opposition to directed change, to inevitable progress, even if you define it in some non-normative way like larger brains and bipedalism. Gould's student Jack Sepkoski put the point forcefully. "I see intelligence as just one of a variety of adaptations among tetrapods for survival. Running fast in a herd while being as dumb as shit, I think, is a very good adaptation for survival" (Ruse 1996, 486). So much for "four legs good, two legs better."

Thus, the basic science. Not that basic value judgments are so very comforting. Take the apparently progress-confirming effects of agriculture and industrialization, certainly something that gives the collywobbles to someone approaching progress from a mechanistic perspective, where the cultural/biological analogy is all-important. In *Homo sapiens* we have a species with an ongoing population explosion, an uneven distribution of wealth, living in a self-polluted environment, and, thanks to its superior intelligence, having produced weapons of horrendous mass destruction. Someone who thinks we have something of absolute supreme value is more credulous than I. In fact, here biology takes back over from value judgments. Short-term gains can be disastrous in the long run. Does anyone truly think that, let us say in the next 20,000 years – a tenth of the as-yet life span of *Homo sapiens sapiens* – no rogue country or determined group of religious or political fanatics will ever obtain and let off a nuclear weapon? Can they guarantee that there will be no event bringing on world-wide annihilation, Dr. Strangelove style? Suppose Hitler's scientists had handed him the bomb in February 1945. Does anyone doubt that he would have at once dropped it on Britain, Russia, and the United States of America?

Complexity

And yet! The conviction persists that we are superior. As G. G. Simpson (1964) used to say, if other organisms dis-agree with this judgment, let them speak up and make their cases. Perhaps, Gould and other skeptics notwithstanding, thanks to evolution, there is some mark of progress that does keep increasing through time. Go back to Clifford and his argument for panpsychism.

> The only thing that we can come to, if we accept the doctrine of evolution at all, is that even in the very lowest organism, even in the Amoeba which swims about in our own blood, there is something or other, inconceivably simple to us, which is of the same nature with our own consciousness, although not of the same complexity (Clifford [1874] 1901, 38–39.)

Complexity! Could this be the mark of real – absolute – progress? In an early notebook, we see Darwin himself toying with this line of thought. He wrote:

> The enormous number of animals in the world depends of their varied structure & complexity. – hence as the

forms became complicated, they opened fresh means of
adding to their complexity. – but yet there is no <u>necessary</u>
tendency in the simple animals to become complicated
although all perhaps will have done so from the new
relations caused by the advancing complexity of others. –
It may be said, why should there not be at any time as
many species tending to dis-development (some
probably always have done so, as the simplest fish), my
answer is because, if we begin with the simplest forms &
suppose them to have changed, their very changes ~~ton~~
tend to give rise to others. (Darwin 1987, E 95–97)

Immediately after the just-quoted passage, Darwin added:
"it is quite clear that a large part of the complexity of
structure is adaptation." Darwinian evolution leads to complexity. Humans are the most complex. We won and were
bound to win.

Not quite so quickly. For a start, you try pinning
down what is meant by "complex." It all seems so easy. *The
Critique of Pure Reason* is complex. A speech by Donald
Trump is not. But go on. Try doing it with organisms. Am
I truly more complex than my cairn terriers, complex, that
is, in a way that makes some biological sense? I am smarter
than they are, sometimes. They are faster than I am, always.
Paleontologist Daniel McShea (1996) gives a thoughtful and
intuitive discussion of the issue. Most obviously, one
simply counts parts and, the greater the number, the more
complex. In the accompanying figure, A is more complex
than B (Figure 15). McShea calls this "object complexity"
(479). Richard Dawkins (2003) endorses something along
these lines. Comparing a lobster with a millipede, what are

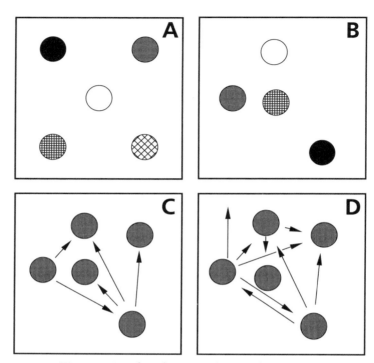

Figure 15 A and B, object complexity. A has more elements or parts than B, and hence is more complex. C and D, process complexity. D has more connections than C, and hence is more complex. (Reproduced by permission of Dan McShea.)

the comparative lengths of books describing them at the same level of detail? "Divide the word-count in the one book by the word-count in the other, and you will have an approximate estimate of the relative information content of lobster and millipede" (100). The trouble is that one runs straight into the problem that bedeviled phenetic taxonomy. What counts as a unit of comparative level of detail? Compare humans with cairn terriers. "And even the very hairs of your head are all numbered" (Matthew 10:30). Probably Dawkins is less than convinced by what Jesus tells us, but I am not sure the rest of us should be so insouciant. Start counting the hairs on the body of Michael Ruse and the hairs on the body of Scruffy McGruff. I am going to be behind from the beginning. Suppose you say, in a hearty voice – "One head, one feature. Are you bald or are you hairy?" Well, okay. What about "One set of cells, one feature. Do you have cells in your body?" Does this mean an amoeba is as complex as a gorilla?

In any case, even if it worked, it doesn't seem quite to capture what we mean by complexity. Compare two calculating machines. The one can do addition. The other can add, subtract, multiply, and divide, and do these all as part of the same calculation. The second is more complex, not just in number of parts, but in interactions between parts. Figure 15, C and D, illustrates this. D has what McShea (1996) calls greater "process complexity" (479). And this does seem to capture intuitions. My brain can do a lot of the things that my cairns can do, but I have more functioning, interacting parts – ability to speak and to do mathematics, for example. More than this, it does seem that life's history has seen an increase in such complexity. Dawkins is right about this. "Once there was only blue–green slime and now there are sharp-eyed metazoa" – and sharp-eyed metazoa are more complex than blue–green slime.

Unfortunately, it is not yet time to declare victory. Indeed, McShea cautions that there may be no victory.

Is a human more complex than a trilobite overall? The question seems unanswerable in principle because the types of complexity are conceptually independent. The aspects of other measures, such as size, have this same independence: a balloon can be larger than a cannonball in volume but smaller in mass. Likewise, a trilobite might have fewer parts but more interactions among parts. Thus, it is hard to imagine how a useful notion of overall complexity could be devised. (McShea 1996, 480)

Adding to the woes, even if there is a measure of complexity, this doesn't mean that things are necessarily more desirable – which is a mark of genuine progress – or that Darwinism promotes such complexity-increase. I like big brains and I like doing philosophy over being dumb as shit and running around in a herd – I am not a football player – but is this necessarily a good evolutionary move? I fear we come up against the Sepkoski problem. Big brains are all very well, but they are very expensive to maintain. We prefer big brains, no doubt, but whether nature feels the same way is another matter. The coronavirus is on a roll at the moment, whereas the chimpanzee in its natural habitat is deeply threatened.

Is this a surprise to any Darwinian? In the Darwinian world, it is not complexity as such that counts, but complexity as superior adaptation. As McShea (1991) points out, this just isn't necessarily true. Go back to Owen's vertebrate archetype (Figure 10). The vertebral column of the marine animal is significantly simpler than those of the land vertebrates, not to mention the birds and the humans. But, for a life in the deep, it is just the job. Can anyone say that the blue whale, coming in at up to 200 tons, able to dive at least half a kilometer, swim at over 30 kilometers per hour, live up to ninety years, is not superbly adapted? It is true that humans with their upright backs can track down and exhaust prey, but we all know that human vertebral columns are far from perfect. They are, I suppose, great adaptations for the bank accounts of chiropractors and orthopedic surgeons.

Why Progress? Arms Races

Remember that the task of this chapter is to show that Darwinism, a mechanistic theory, can yield biological progress. Complexity, I am afraid, raises more problems than it solves. Go at the problem another way. Can we show in some sense that evolution does promote organisms with ever-better adaptations? There are at least three proposed strategies. The first takes us right back to the heart of the cultural Progress/biological progress analogy. It centers on the notion of arms races, where lines of organisms compete against each other, thereby improving their adaptations. The prey gets faster and in tandem the predator gets faster. At the end of the last chapter we saw how Darwin floated a kind of proto-version of this argument. This military-analogy idea was developed in detail by Julian Huxley in his first book, *The Individual in the Animal Kingdom* (1912). First, the cultural side, using naval warfare as the example. "Halfway through the century, when guns had doubled and trebled their projectile capacity, up sprang the 'Merrimac' and the 'Monitor,' secure in their iron breastplates; and so the duel has gone on" (115). Concluding: "Each advance in attack has brought forth, as if by magic, a corresponding advance in defence." Then, the biological: "With life it has been the same: if one species happens to vary in the direction of greater independence, the inter-related equilibrium is upset, and cannot be restored until a number of competing species have either given way to the increased pressure and become extinct, or else have answered pressure with pressure" (115). Adding: "So it comes to pass that the continuous change which is passing through the organic world appears as a succession of phases of equilibrium, each one on a higher average plane of inde-pendence than the one before, and each inevitably calling up and giving place to one still higher" (116).

Jumping to the present, Richard Dawkins goes right along with this kind of thinking. He brings up the increasing employment by competing nations of ever more sophisti-cated computer technology. In the animal world, Dawkins sees the evolution of bigger and bigger brains. As with Huxley's independence, so with Dawkins's brain power. We won! In Chapter 1, we saw how human beings have wrestled the IQ-challenged hippopotamus to the ground.

Why Progress? Convergence

The second progress-producing process centers on conver-gence. The key starting point is that of the niche, a space that organisms occupy. Fish, for instance, have colonized the ocean niche. Birds, the air-above-us niche. British paleon-tologist Simon Conway Morris (2003) argues that only cer-tain areas of potential morphological space will be able to support functional life. This puts constraints on the direction of evolution. Not all pathways are open. But those that are open tend to be used again and again – "convergence." The same adaptive morphological space is shared by different occupants. The most dramatic and well-known case is that of saber-toothed-tiger-like organisms, where the North American placental mammals (real cats) were matched right down the line by South American marsupials (Figure 16).

From examples like this, Conway Morris concludes that the historical course of nature is not random but strongly selection-constrained along certain pathways and to certain destinations. In other words, life's history was far from haphazard as someone like Sepkoski rather implies. And this opens the way for progress. It happened and had to happen. Come what may, sooner or later some kind of intelligent being (a "humanoid") was bound to emerge. After all, our own very existence shows that a kind of cultural adaptive niche exists – a niche that prizes intelligence and social abilities. It was waiting for us and we grabbed it! Humans came into their own. It was their Darwinian destiny! "We may be unique, but paradoxically those properties that define our uniqueness can still be inherent in the evolutionary process. In other words, if we humans had not evolved then something more-or-less identical would have emerged sooner or later" (Conway Morris 2003, 196).

Why Progress? Complexity

Turn to a third and final attempt to get progress from, or despite, the Darwinian process. This takes us back to the

Placental Sabertooth

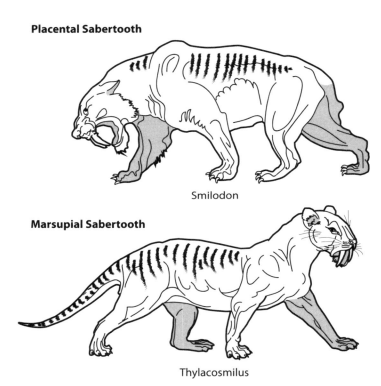

Smilodon

Marsupial Sabertooth

Thylacosmilus

Figure 16 Placental and marsupial sabertooth tigers compared. Technically, the Thylacosmilus is a sparassodont, a group closely related to marsupials.

notion of complexity. Somehow – as Darwin rather implies in notebook musings – it just emerges because, in a way, you are always going to be building on what you have already. Earlier, we saw Darwin toying with this strategy: "The enormous number of animals in the world depends on their varied structure & complexity" (Darwin 1987, E 95). We saw also that natural selection was not left entirely out of the mix, or, perhaps more accurately, we might say that Archdeacon Paley was not left entirely out of the mix: "it is quite clear that a large part of the complexity of structure is adaptation."

Others, more recently, have taken up the case. Somewhat surprisingly given his earlier thinking, the already-encountered paleontologist Daniel McShea, backed by his colleague at Duke University, philosopher Robert Brandon, sees in life's history the potential for just that kind of upwards momentum to life's history that Darwin toyed with. Favoring a somewhat inclusive notion of complexity, in terms of number of part types, they introduce what they call the "zero-force evolutionary law" (ZFEL for short). They write: "In any evolutionary system in which there is variation and heredity, in the absence of natural selection, other forces, and constraints acting on diversity or complexity, diversity and complexity will increase on average" (McShea and Brandon 2010, 3). It seems that for them things just naturally keep complexifying – one cause leads to several effects and these in turn multiply. Thus, all falls into place.

Why the Enthusiasm?

What do we say to all of this? Gould is right, Sepkoski even more so. Darwinism is not friendly to biological progress. There must be something fishy going on. Fried kippers have nothing on the pong around the author of the *Origin*. Say it again. Darwin was a great revolutionary. He was no rebel. Cultural Progress was the foundation of Darwin's personal life. He and his wife – a first cousin – were grandchildren of the hugely successful, late-eighteenth-century potter Josiah Wedgwood, hence embedded in the comforts and security of an industry-based, Silicon Valley-size, family fortune, citizens of the country that was painting a quarter of the globe red. Darwin was going to find progress in his theory, a move eased by the fact that, as a teenager, he read his other grandfather's progress-impregnated speculations on evolution. True, Darwin knew the problems, as reflected by his notebook comment – "there is no necessary tendency in the simple animals to become complicated" (E 95, January 1839). But, he had to get it somehow, so one is really not surprised by the progressivist sentiments of the *Origin* or the proto-arms-race speculations that came shortly after that book was published. Darwin is a mechanist, but this is seen through the lens of his culture, family and societal.

Julian Huxley was rather differently motivated. More than one person has remarked on the paradox that, although Huxley was the great Darwinian synthesizer, his underlying commitment to natural selection was less than enthusiastic, at best. Huxley did not get his progressivist thinking from Darwin. He got it from the French philosopher Henri Bergson, whose *L'évolution créatrice* (1907) greatly influenced Huxley (he was fluent in French) and was completely the underlying philosophy of *The Individual in the Animal Kingdom*. Bergson's enthusiasm for Spencer proved infectious. Remember: "the organic world appears as a succession of phases of equilibrium, each one on a higher average plane of independence than the one before" (Huxley 1912, 116). From Spencer back to Schelling and a philosophy of nature rising teleologically to the top. No wonder, to the Christian

Dobzhansky's horror, atheist Julian Huxley was president of the British branch of the Teilhard de Chardin Society.

Suggestively, Edward O. Wilson, fanatical progressionist, shows no great need to justify his position. For all his public Darwinism, privately Wilson obeys different masters (Gibson 2013). On my expressing surprise that he had, on his laboratory wall, a larger and more prominent picture of Spencer than of Darwin, he responded: "Great man, Mike, great man!" Wilson is a product of the Harvard biology department, already noted as a unit with a long history of enthusiasm for the holistic, neo-Romantic thinking of Herbert Spencer. Wilson truly is into thinking of groups as super-organisms. The "comparison of colonies and organisms has a larger goal than amusing analogy: the meshing of comparable information from developmental biology and sociobiology reveals more general and exact principles of biological organization" (Wilson 1990, 57). In the hymenoptera, the individual insects are parts of the whole, with no expectations that the parts will be similar. The queen is different from the workers and the workers are different from the drones. Going beyond kin groups, almost uniquely among leading evolutionists, Wilson is an enthusiast for group selection. Like all Romantic thinkers, biological progress simply comes with the territory. No need to worry about causes. As we consciously strive to ends, so nature likewise has an inherent force striving for ends. It is practically gilding the lily to point out that Wilson's "pinnacles" bear a suspicious resemblance to Spencerian "equilibria." We get to one level, take a breather, and try again to get to a higher level – humans at the top.

As with Spencer, with Huxley and Wilson there is more than a faint odor of organicism. By contrast, no one could question Richard Dawkins's wholehearted commitment to Darwinian evolutionary theory, in all its mechanistic glory. With the *Selfish Gene* he wrote the book on it! Charitably, one might say, he has not really thought through the full implications of his beliefs in progress. Less charitably, one might say, such surface thinking is not unknown in the secular halls of New Atheism. There is a touch of the Book of Job about the arguments of that crew. What they say is true because they said it. "Where were you when I laid the earth's foundation? Tell me, if you understand" (38:4). It never occurs to any of them that they might not be the apotheosis of a teleologically driven process of evolution. Cultural Progress leading up to them; ergo, biological progress leading up to them. At least one can say that Simon Conway Morris is a little more open in this prejudice. He is a conservative Christian! One suspects here an interesting mélange of Providential and Progressionist thinking. As a Christian, he knows that humans were bound to appear. As a scientist, he wanted a law-bound way in which this could occur. Hence, the convergence argument. Without being overly cynical, one suspects the scientist is the handmaiden of the Christian.

With McShea and Brandon, we are right back in the nineteenth century: "we identify the intellectual ancestor of our view as Herbert Spencer." Whatever else, they have the same self-regard for their accomplishments as had that Victorian bachelor long ago.

> The scope we claim for the ZFEL is immodestly large.
> The claim is that the ZFEL tendency is and has been
> present in the background, pushing diversity and
> complexity upward, in all populations, in all taxa, in all
> organisms, in all timescales, over the entire history of life,
> here on Earth and elsewhere. (McShea and Brandon
> 2010, 134)

Not a lot of Darwinian selection to be seen here. And, of course, there isn't any. The paradox of McShea is answered. How, at one point, he can be so skeptical about complexity and progress and, at another point, endorse the link. When he is considering the problem from a mechanistic, Darwinian perspective, where adaptation is all-important, complexity does not equal progress. When he is considering the problem from an organicist, Spencerian perspective, where adaptation is secondary, complexity does equal progress.

Causal Inadequacies

Natural selection simply does not guarantee winners and losers, and so you look with a pinch of salt at the arguments for biological progress in a Darwinian world – a world of the machine. Start with arms races. They seem plausible. The prey gets faster, the predator gets faster. Note, though, that this is comparative progress, which is not normative. Even if arms races can produce comparative progress, you have still got to show that they occur and that they are reasonably frequent, if not the norm. There are some genuine cases. Arms races between shellfish and their predators are well documented. The predators develop ever-better methods of boring into the shells. The shells get ever thicker and harder to penetrate (Dietl 2003). This said, not every evolutionist thinks that arms races are all that prevalent. Paleontologist Robert Bakker (1983) is skeptical about mammalian arms races between predators and prey. In any case, unless you are convinced that what we know of shellfish evolution translates directly into the necessary evolution of human beings, the peak of biological perfection, the support of the arms-race hypothesis seems less than overwhelming. One has a nasty feeling that, although Dawkins trumpets that it is comparative progress he is after, truly, with his emphasis on humans and brains, hopes of absolute progress lurk.

The same seems true of Conway Morris's claims about niche occupation. Many biologists would challenge the way in which Conway Morris conceives niches. He thinks of them as objective things, "out there," waiting to be occupied. Throwing cold water on this belief, Richard Lewontin writes: "organisms not only determine what elements of the outside world are relevant to them by peculiarities of their shapes and metabolism, but they actually construct, in the literal sense of the word, a world around themselves" (Lewontin 2002, 54). This is a biological worry. And now a theological worry: Is Conway Morris making a case too strong for its own good? Remember J. B. S. Haldane. Not only is the world queerer than we think it is, but it is queerer than we can think it is. Who is to say that there are not niches out there waiting for beings to evolve to the point where they can enter and achieve a higher state of being? Beings to whom we humans are but children – perhaps warthogs? Dizzying thought that perhaps someone might so regard my late headmaster. Perhaps not. Warthogs have redeeming features.

I am not sure that anything at all points positively towards the thinking of McShea and Brandon. How you get humans out of ZFEL is an even greater mystery than the mind–body problem. The fascinating thing is that McShea and Brandon themselves seem not to have the courage of their convictions. Having introduced ZFEL with its claims about complexity, they take it all back!

> We conclude what is known about the history of life offers little evidence for the ZFEL for complexity. A long-term increase in the mean has not been demonstrated, but if in fact it occurred, it would be consistent with a number of possible underlying mechanisms. The ZFEL predicts a strong drive, but no such drive has been shown, and indeed the stable minimum suggested by impressionistic assessments argues for the opposite, a weak drive or none at all. (84)

I am put in mind of Mr Micawber in *David Copperfield*. He thought, by writing out IOUs for all his debts, he had solved his financial problems. McShea and Brandon think that, by listing the problems with their position, they have validated it.

Values

What do we conclude? Simply this. Darwin and, assuming that this is what they believe, McShea and Brandon were right that we do see an increase in some notion of complexity. This is because, if you start at the bottom, there is nowhere to go but up. It is surely the case that natural selection, through arms races and the like, had a hand in this. Darwinian selection played a major role in driving the evolution of humans up from the other great apes. However, this is not to say that, in any overall sense, Darwinian selection promotes progress. The Sepkoski objection stands. Nor is it to say that the most complex are necessarily the peak of Darwinian desirability – think nuclear weapons and our future. It is true that life today for some humans is much better than was the life of the peasant or serf in the Middle Ages. But that is a value judgment that we make, not Darwinism. One suspects many in the world today would not make such a judgment. Mechanism simply cannot do the job.

All of this said, there is obviously still a powerful drive to see humans as the evolutionary success. Why? Simply, we are caught in one of those paradoxes like Descartes's *cogito ergo sum*. To deny it is to affirm it. To deny progress to humans is (as Simpson said) to use those very attributes missing in others. To deny it is to affirm it. Yet, given the objections to progress, although this Cartesian-style argument may be prima facie convincing, it cannot be valid. You are assuming reason and language ability important, which is the very thing you are trying to prove. Articulate or not, Sepkoski's tetrapods might have a different take on things. Ability to find fodder in a calorie-deprived area for instance. Darwin's theory just does not guarantee (absolute) progress. And, expectedly and revealingly, the more you seem to get a guarantee of progress from science, as from Julian Huxley or Simon Conway Morris or Edward O. Wilson or Dan McShea and Robert Brandon, the less you seem Darwinian. The more you seem

148

to be moving from the machine metaphor to the organism metaphor.

I suppose that, even in a Darwinian world, if you believe in an infinite number of multiverses, human evolution was bound to happen, an infinite number of times. Paradoxically, Gould (1985) believed something of this nature: "I can present a good argument from 'evolutionary theory' against the repetition of anything like a human body elsewhere; I cannot extend it to the general proposition that intelligence in some form might pervade the universe." Unfortunately, you are also going to have infinite numbers of also-rans – would-be humans with the IQs of turnips and the sporting ability of Michael Ruse. I do not know whether in such a case one would want to speak of progress. So I am not sure you solve a theological problem we have encountered – how, given the non-directionality of the Darwinian process, we can reconcile Darwinism with the underlying Christian assumption that, given God's creative powers, humans were bound to appear. I doubt God wants multiverses full of turnip-IQ, soccer-playing Ruses. I suppose you might argue that God, seeing the possibilities, picked the one he wanted. Like Dobzhansky, I worry a bit about this. We don't really have directed evolution, at least not as we normally think of directed evolution. Yet, if quantum processes really are random, I am not sure that beforehand even God could tell which possibility would produce humans. Just that some will do so.

Before the organicist gets too smug about the tangles of the mechanist with Christian doctrine, I might point out that it is all very well making so much of progress.

One is liable to forget that unadorned Progress is heretically opposed to Providence. Biological progress is not cultural Progress, and the analogy is not central for the organicist, but they are still Siamese twins. Take care! For now, leave things. Darwinism? No guaranteed biological progress. We really knew this before we started. Progress is a value-impregnated notion. Darwinism eschews (absolute) values. Never the two shall meet.

7   Morality for the Organicist

I will give first the organicist position, and then turn to the mechanist position. I will put to one side (but not entirely ignore) views of morality, like Platonism, that discount human nature, especially biological human nature. You have still got to show how creatures like human beings might be expected to grasp these empyrean views.

Two Divisions

I start with two divisions, already introduced. The first, and I take it to be entirely standard, is the already-mentioned one made in the philosophy of morality, ethics, between what are known as "normative" or "substantive" claims and what are known as "metaethical" claims. The former refer to prescriptions for thought and behavior. The latter refer to justifications of prescriptions for thought and behavior, that is, justifications of normative or substantive claims. Take, by way of illustration, morality in the Christian system. On the one hand, there are prescriptions for action. Be kind to strangers. Help the sick and the poor. Show concern for those imprisoned. On the other hand, Christians have justifications. Often the justification is based on the Will of God. A familiar objection is the so-called Euthyphro Problem, named after Plato's dialogue of that name. Should I do right because it is the Will of God, or is the Will of God that which is right? Could God make it okay to mark up library books with yellow felt pens? Sophisticated Christians have answers to these questions, for instance the Catholic doctrine of natural law. We must do what God wants us to do but he wants us to do that which is natural. Sodomizing children is unnatural because that goes against the way we are made. Hence, it is wrong. God designed and made the world; but, given that design, morality (at the substantive level) emerges automatically.

Any system of morality based on, or in some way reflecting, a theory of evolution must pay attention both to normative or substantive ethics and to metaethics. What should the evolutionist think and do? Why should the evolutionist thus think and do? These are not trick or unfair demands. They must be spoken to. And getting the discussion under way brings in our second division – our different root metaphors. The older root metaphor is that of an organism. The world is to be regarded organically. The newer root metaphor is that of a machine. The world is to be regarded mechanistically. Darwinian theory is mechanistic. I say this notwithstanding that, primarily because of his culture, particularly his own background, Darwin was not always entirely true to his philosophy. He gets results – progress – that no true Darwinian should get. Parenthetically, I must acknowledge that the eminent Chicago historian of science, Robert J. Richards, sees things otherwise. He stresses the influence on the young Darwin of the explorer and travel writer Alexander von Humboldt, and has gone so far as to say that Darwin bought in entirely to the Romantic philosophy. "In Darwin's view, the world and all of its organisms were made for man" (Richards and Ruse 2016, 122). Appearances to the contrary, Darwin's true roots were in Germanic biology and thought generally.

Take your choice. False to mechanism (Ruse). Loyal to organicism (Richards). Either way – progress! Our differences fade beside the agreement that this is at the heart of the organic model. The organism is born, grows, changes until it reaches full blossoming. What about subsequent decay and death? This always hovers and is a threat. Diderot, remember, was more into the P/progress analogy than into organicism for its own sake, so he could be candid: "an individual begins, so to speak, grows, subsists, decays and passes away." But, for the true organicist, that side to organisms is cut out from the metaphor as purpose is cut out of the machine metaphor. We have a positive-value-impregnated picture. We have a holistic picture. We have a progressive picture. Simple to complex. Value-free to value-laden. Monad to man.

Social Darwinism

We have therefore two possible approaches to evolution and morality – through the organic metaphor and through the machine metaphor. It makes sense to take up the organic first, the Romantic interpretation of evolution, for it is this that has dominated discussions of evolution-influenced moral thinking. Little surprise, one supposes, because it is this metaphor rather than the machine metaphor that stresses values. One suspects that Darwinism haters like John Dupré and Jerry Fodor would not be entirely grateful to Richards for his energetic argument that the organic paradigm is truly the paradigm of Darwin. Be this as it may, an organicist ethics has more often than not been regarded as a direct outcome of the evolutionary stance of the *Origin of Species*, to the extent that, increasingly, it was known generally as "Social Darwinism" (O'Connell and Ruse 2021).

At once you may recoil, hoping that we can bring this part of the discussion to a rapid end. If anything has a bad reputation, it is Social Darwinism. The worst kind of laissez-faire, sociopolitical theory, beloved of the harshest industrialists. Widows and children to the wall. I'm alright Jack! Listen to Herbert Spencer. "We must call those spurious philanthropists, who, to prevent present misery, would entail greater misery upon future generations. All defenders of a Poor Law must, however, be classed among such. That rigorous necessity which, when allowed to act on them, becomes so sharp a spur to the lazy and so strong a bridle to the random, these pauper's friends would repeal, because of the wailing it here and there produces" (Spencer 1851, 323). When you learn that there is a direct line from this kind of thinking to the vile policies of the Third Reich, little wonder that Social Darwinism is regarded with horror and rejected by all but the truly unnatural. But before we pack up, bringing this discussion to a rapid end and on into the Index, let us pause and ask a couple of pertinent questions. First, is this truly Darwinian? Is there truly a clear route from the Charles Darwin we have encountered thus far in these pages – the important Darwin, revolutionary not rebel – to the archfiend of the twentieth century? Second, is this an entirely fair characterization of the movement? For the moment, we can postpone the third pertinent question, what is the purported metaethical justification for this kind of stuff?

There is a common tendency to exonerate Darwin entirely and to lay it all at the feet of Herbert Spencer. However, especially given his organicist-like yearnings, it would be naïve to predict that no such views are to be found in Darwin's writings. From a letter written almost at the end of his life: "The more civilised so-called Caucasian races have beaten the Turkish hollow in the struggle for existence. Looking to the world at no very distant date, what an endless number of the lower races will have been eliminated by the higher civilised races throughout the world" (To William Graham, July 3, 1881, Darwin Correspondence Project, Letter 13230). This is but one side of the story. Famous was the Darwin family hatred of slavery. Likewise, in the political world, Darwin's hope was for the (selection-driven) end of war. "As man advances in civilisation, and small tribes are united into larger communities, the simplest reason would tell each individual that he ought to extend his social instincts and sympathies to all the members of the same nation, though personally unknown to him. This point being once reached, there is only an artificial barrier to prevent his sympathies extending to the men of all nations and races" (Darwin 1874, 122).

The story is mixed, but Darwin was certainly no rabid Social Darwinian. Nor, to answer the second question, were most Social Darwinians! You can certainly find such sentiments if you look. Andrew Carnegie, founder of US Steel, one of the most successful and powerful of the late-nineteenth-century American titans of industry, can readily outdo Spencer when it comes to his harsh views on society. "The law of competition may be sometimes hard for the individual, [but] it is best for the race, because it insures the survival of the fittest in every department" (Carnegie 1889, 655). No Anglophone, however, can outdo German militarists. General Friedrich von Bernhardi, a sometime member of the German General Staff, was unequivocal. "War is a biological necessity," and hence: "Those forms survive which are able to procure themselves the most favourable conditions of life, and to assert themselves in the universal economy of nature. The weaker succumb" (von Bernhardi 1912, 10).

There is another side to the story. The passage by Spencer quoted above, written before he became an evolutionist, is directed less against the poor and more against the rich and powerful who grab all of life's goodies. Anticipating Margaret Thatcher, a century later – like Spencer from the non-conformist, provincial (British Midlands), lower middle classes, hating the parasitic rich – Spencer wanted fewer rules so the poor-but-merited could rise in society. Slackers must not be coddled, but they must be given the chance. Later in life, as he started to push the integrated nature of society, he fulminated against German militarism which, among other things, he saw as bad for trade. Andrew Carnegie (1889) likewise had a different side. He is better known – and rightly so – for his philanthropy. Use your money for the good of society. "Under its sway we shall have an ideal state, in which the surplus wealth of the few will become, in the best sense the property of the many, because administered for the common good, and this wealth, passing through the hands of the few, can be made a much more potent force for the elevation of our race than if it had been distributed in small sums to the people themselves" (655). His life's mission was the sponsorship of public libraries. These are places where the ambitious and talented can learn and move up in society, benefiting themselves and all around them.

Nothing will ameliorate the views of the German militarists. But, apart from the fact that von Bernhardi loathed and detested the British – hardly what one expects from a Darwin disciple – home-grown philosophies had a greater hold on their imaginations. "Life merely as such, the mere continuance of changing existence, has in any case never had any value for him; he has wished for it only as the source of what is permanent. But this permanence is promised to him only by the continuous and independent existence of his nation. In order to save his nation, he must be ready even to die that it may live, and that he may live in it the only life for which he has ever wished" (Fichte 1821, 136). Much the same sentiments apply to Hitler and his gang. For all that there are Darwinian-sounding sentiments, whether the thinking is ever truly Darwinian may be queried. Hitler himself didn't believe in evolution! And he certainly wasn't going to be keen on any tree of life that linked Aryans and Jews (Richards 2013).

Whatever its reputation, this doesn't mean that the organicist approach to morality was relinquished. Many, biologists particularly, have found moral prescriptions in the evolutionary process. Julian Huxley wrote enthusiastically on the relationship between evolution and ethics.

He argued that evolution justifies an obsession with technology, science, and major public works: "the individual is meaningless in isolation, and the possibilities of development and self-realization open to him are conditioned and limited by the nature of the social organization. The individual thus has duties and responsibilities as well as rights and privileges, or if you prefer it, finds certain outlets and satisfactions (such as devotion to a cause, or participation in a joint enterprise) only in relation to the type of society in which he lives" (Huxley 1934, 138–39).

Today there is Edward O. Wilson (1975) – "scientists and humanists should consider together the possibility that the time has come for ethics to be removed temporarily from the hands of the philosophers and biologicized" (3). You can imagine how this went down in the analytic philosophical community. Not that this was about to stop Wilson. As Julian Huxley's prescriptions reflected the challenges of his era, so Edward O. Wilson's prescriptions reflect the challenges of our era. Wilson has concern about the environment, specifically about biodiversity (Wilson 1984, 1992, 2012). This is expressed through his "biophilia" hypothesis. "To explore and affiliate with life is a deep and complicated process in mental development. To an extent still undervalued in philosophy and religion, our existence depends on this propensity, our spirit is woven from it, hope rises on its currents" (Wilson 1984, 1). In the organicist tradition, he sees all of life as an interconnected whole. Individual organisms, individual species, are part of a larger network, and no one or group can take itself apart in isolation. Morally, therefore, our obligation is to preserve life.

The Naturalistic Fallacy

At the normative or substantive level, this approach to evolutionary ethics – Social Darwinism, or whatever you would call it – is more complex, interesting, and often more morally defensible than popular lore would have it. What of its metaethical foundations? With enthusiastic exponents like Herbert Spencer, Julian Huxley, and Edward O. Wilson, not to mention the German militarists, there is not much surprise. The foundations lie in the supposedly progressive nature of the evolutionary process, with humans at the top. Darwin's dithering aside, the thinkers we have considered in this chapter are into progress in a very big way. Since nature is an organic unfolding, getting ever more perfect, prescriptions emerge naturally. We ought to cherish the evolutionary process as generating ever greater value, and hence we ought to help it along. At least, not impede its progress. "Ethics has for its subject-matter, that form which universal conduct assumes during the last stages of its evolution" (Spencer 1879, 21). Adding: "And there has followed the corollary that conduct gains ethical sanction in proportion as the activities, becoming less and less militant and more and more industrial, are such as do not necessitate mutual injury or hindrance, but consist with, and are furthered by, co-operation and mutual aid."

For von Bernhardi, there is progress and it depends on war: "Without war, inferior or decaying races would easily choke the growth of healthy budding elements, and a universal decadence would follow" (von Bernhardi 1912, 20). In a very different key, Julian Huxley sings the same song:

> I do not feel that we should use the word purpose save
> where we know that a conscious aim is involved; but we
> can say that this is the most desirable direction of
> evolution, and accordingly that our ethical standards
> must fit into its dynamic framework. In other words, it is
> ethically right to aim at whatever will promote the
> increasingly full realization of increasingly higher values.
> (Huxley 1927, 137)

Expectedly, the ethical "biologicizing" Edward O. Wilson is
right into this sort of stuff.

> Human beings face incredible social problems, primarily
> because their biology cannot cope with the effects of their
> technology. A deeper understanding of this biology is
> surely a first step towards solving some of these pressing
> worries. Seeing morality for what it is, a legacy of
> evolution rather than a reflection of eternal, divinely
> inspired verities, is part of this understanding. (Ruse and
> Wilson 1985, 52)

The whole point of the biophilia hypothesis is to preserve
and further the nature and fortunes of humankind.

The traditional philosophical objection to this kind
of approach is that you cannot go from "is" to "ought." You
cannot go from the way evolution occurs – a fact of nature –
to this is what evolution tells you you should do – a moral
prescription. The objection goes back to Hume. We should
not slide without notice from talking about what is, to what
ought to be. "For as this ought, or ought not, expresses some
new relation or affirmation, 'tis necessary that it should be
observed and explained; and at the same time that a reason
should be given, for what seems altogether inconceivable,

how this new relation can be a deduction from others, which are entirely different from it" (Hume 1739–40, 302). Post-Darwinian philosophers made this their trademark objection to evolutionary ethics. Henry Sidgwick, in the first year of the journal *Mind*, stated bluntly that "the theory of Evolution, thus widely understood, has little or no bearing upon ethics" (Sidgwick 1876, 56). Spencer is the focus of attack, as he is in the much-celebrated book *Principia Ethica*, by Sidgwick's student G. E. Moore. Spencer's attempt to get morality out of evolution is taken to be an egregious example of someone committing what Moore labeled the "naturalistic fallacy," trying to get ought from is: "he tells us that one of the things it has proved is that conduct gains ethical sanction in proportion as it displays certain characteristics. What he has tried to prove is only that, in proportion as it displays those characteristics, it is more evolved" (Moore 1903, 31). Alas, "more evolved" is a matter of fact. "Conduct gains ethical sanction" is a matter of obligation. You cannot legitimately go from the one to the other. The pattern was set for the rest of the twentieth century. Moore's student C. D. Broad (1944) in turn berated Julian Huxley. By the time I started doing philosophy in the 1960s, the absoluteness of the naturalistic fallacy was one of the eternal verities, along with "even the nicest boys want only one thing" and "Americans have terrible table manners." (Cut up the meat first and then eat it with a fork?) I still believe in it, as I still suspect the desires of the nicest boys – I am no longer quite so sure about American table manners, but in these days of fast food who ever uses a knife and fork?

End of discussion. Well, yes, but don't go too quickly. To revert to the conclusion at the end of the last chapter, it is all a question of values. The progressionist says that the end point of evolution is better than the beginning: blob to human. This is totally in harmony with the organic metaphor. The developed organism – oak tree, butterfly, adult human – is superior to the acorn, the caterpillar, the fertilized ovum. Values are built into the perspective. All these evolutionary ethicists subscribe to aspects of Romantic thinking. Spencer and his debt to Schelling. Huxley and his debt to Spencer directly and indirectly through Bergson. Wilson likewise indebted to Spencer. And von Bernhardi had the whole range of Germanic thinking open to him. Thus Hegel: "Nature is to be regarded as a system of stages, one arising necessarily from the other and being the prox-imate truth of the stage from which it results" (1817, 20). Values come with the territory, or, less metaphorically, with the metaphor!

One of the striking things about paradigms or root metaphors or metaphysical world-pictures is how difficult it can be to see the point of the other side. I trust the reader's interest was piqued by the fact that I quoted a passage by Wilson of which I was co-author. To put it in context, I shared with Wilson the conviction that evolution was important. But whereas Wilson saw biology justifying morality, I saw it giving a genetic account of morality. Clarifying, I went on explicitly to develop a kind of neo-Humean position (Ruse 1986) – the topic of the next chapter. It was not so much that Wilson did not accept the is/ought distinction, the naturalistic fallacy, but that he simply could not see it! For him, it was not so much wrong as invisible. That is the mark of paradigm-thinking, and his paradigm was not my paradigm.

The point I am making is not whether one should accept the Humean is/ought distinction. I do. Rather, you are not going to cut much ice by waving it in the organicist's face. They just don't accept that way of thinking. This is not the end of the argument, quite apart from general arguments in favor of mechanism over organicism. The organicist has still to explain how, if what has evolved is good, some things that have evolved in humans, and that seem very efficient adaptations, just do not come across at all as very good. We have learnt enough to want to keep this worry in perspective. The essence of being human, if one might so call it, is being social. That is how our ancestors got on and how we get on. Early *Homo sapiens*, for instance, was not into systematic warfare. This seems to have been a function of the coming of agriculture, with the desirable goods that it produces – a temptation to those outside the fold. Add in the ability to make powerful weapons and the like, and you have a recipe for disaster.

> An important milestone in looking at the origins of warfare in humans is 8000 BC, as it stands at the very end of the Mesolithic and the beginning of the Neolithic periods. It also marks major changes in the trajectory of human history as humankind was reaching the upper demographic limits of sustainable hunting and gathering around the world. People were in the throes of the transition from a hunting-and-gathering, nomadic lifestyle to an agricultural and settled lifestyle. (Haas and Piscitelli 2013, 178)

Humans are social. They are not killer apes. But for various reasons they are given to violence and to hostility towards the other – those not members of one's tribe. With the growth of populations and the mixing of people, clearly the intergroup hostility can turn into intra-group hostility. Think of the Nazis and Jews. Think of many today, in the UK and USA, and Muslims. And these are problems for the organicist. One must provide reasons why we might some-times say that violence is natural – fighting Hitler for instance – and why we might sometimes say that aggressive behavior is unnatural – killing people in order to have gay sex with their corpses. The organicist has still got a hard slog ahead.

And this is separate from the issue of the horrible process – vile struggles for survival – that seems required to get all this progress. Someone like von Bernhardi welcomes the struggle. The organicist needs to show why positions like that of Wilson might be welcomed and why positions like von Bernhardi's are perversions of the philosophy, not con-sequences. One possible strategy for the organicist is to argue that, while there are struggles, they are nothing like as pervasive as the Darwinian supposes. Spencer read Malthus and incorporated the arguments into his own pos-ition, but it was less a matter of competition between organ-isms, and more a matter of the stresses of overpopulation leading to personal effort and Lamarckian improvement. A group selectionist like Wilson will tend to see a lot less competition within species. Epigeneticist enthusiasts see no need for struggle. Change is all coming from within. Convincing or not, arguments must be furnished for all these sorts of claims, and I suspect there will be a fair amount of special pleading.

I will not go further with the discussion now. I am not defending the position. Saying simply that, too often, critics simply don't understand the thinking of the organicists. In a brief Epilogue, I will take up the final discussion of how all this fares in light of our three-fold division of Chapter 1.

8   Morality for the Mechanist

Human Moral Nature

How does the machine metaphor deal with morality? Many think there is no way that evolutionary theory – Darwinian evolutionary theory, that is – can lead to morality. Social Darwinism is fatally flawed, and the implication is that the same applies to any other approach. Thomas Henry Huxley is an articulate exponent of this position, one formulated explicitly in opposition to his old friend Herbert Spencer. In his Romanes Lecture of 1893, "Evolution and Ethics," Huxley writes of human evolution: "For his successful progress, throughout the savage state, man has been largely indebted to those qualities which he shares with the ape and the tiger; his exceptional physical organization; his cunning, his sociability, his curiosity, and his imitativeness; his ruthless and ferocious destructiveness when his anger is roused by opposition" (Huxley 1893, 51). However, in a way, this has all been self-defeating, or perhaps more accurately, it means that those things that helped us to our present position are no longer needed, they are in fact hindrances. "Whatever differences of opinion may exist among experts, there is a general consensus that the ape and tiger methods of the struggle for existence are not reconcilable with sound ethical principles."

Lions and tigers and bears, oh my! Stirring stuff and, at one level, one hesitates to disagree. The Thrasymachus view of life is not the correct one. "Might is right" is deeply immoral. Yet, let us not rush to assume that Huxley got Darwin entirely right. He hadn't done so for most of his life, from the *Origin* on, so why assume he did so now? The whole point about humans is that we have taken a very different route from that of apes and tigers – or, more accurately I should say, from the route that the apes and tigers are supposed to have taken, for I am not sure that they are always so very different from us. Natural selection does not always promote conflict and aggression. In the case of humans, we know that this is very much not so. Say it again: the defining part of our nature is that we are social. You don't need to be a Darwinian to know this. The great metaphysical poet John Donne put my thoughts in words to which I can never aspire.

> No man is an island,
> Entire of itself,
> Every man is a piece of the continent,
> A part of the main.
> If a clod be washed away by the sea,
> Europe is the less.
> As well as if a promontory were.
> As well as if a manor of thy friend's
> Or of thine own were:
> Any man's death diminishes me,
> Because I am involved in mankind,
> And therefore never send to know for whom the bell tolls;
> It tolls for thee.
>
> (Meditation 17, *Devotions upon Emergent Occasions*, dated 1624)

Don't ask me. Look at our biology. We get along by getting along. We don't have weapons of attack. I get cheesed off at you. I am unlikely to tear your throat out. We have social instincts. I see a small child, alone and crying in the park. I don't ignore it. I wonder where its parents are. We have adaptations for social life. Hard enough teaching logic as it is. Imagine if three of the class were in heat. Of course, being nice sometimes is hard and annoying. I won't say we always do what we should do – how often has a tired professor skipped out through the back door when a needy, pre-exam student is at the front? – but we know what we should do. Our biology directs us that way. "We have cooperative brains, it seems, because cooperation provides material benefits, biological resources that enable our genes to make more copies of themselves. Out of evolutionary dirt grows the flower of human goodness" (Greene 2013, 65).

There is no magic to any of this. It is very much as it was when we were talking about knowledge. Successful proto-humans took seriously 2 + 2 = 4. Successful proto-humans took seriously the dictate to think of others. Helping this along, selection conferred a sense of self-worth dealing with the needs of others. "When people who are fairly fortunate in their material circumstances don't find sufficient enjoyment to make life valuable to them, this is usually because they care for nobody but themselves" (Mill 1863). All very much in the spirit, almost the literal word, of Charles Darwin in the *Descent of Man*. Diametrically opposed to the Huxley view of humankind, he stressed that sociality is the key to human success. Tribes of people who get along and help each other do better than tribes who don't.

It must not be forgotten that although a high standard of morality gives but a slight or no advantage to each individual man and his children over the other men of the same tribe, yet that an advancement in the standard of morality and an increase in the number of well-endowed men will certainly give an immense advantage to one tribe over another. There can be no doubt that a tribe including many members who, from possessing in a high degree the spirit of patriotism, fidelity, obedience, courage, and sympathy, were always ready to give aid to each other and to sacrifice themselves for the common good, would be victorious over most other tribes; and this would be natural selection. (Darwin 1871, 1, 166)

Darwin speculated on the causes behind this evolution of the moral sense. One suggestion is today known as "reciprocal altruism." You scratch my back and I will scratch yours. He wrote: "as the reasoning powers and foresight of the members [of a tribe] became improved, each man would soon learn from experience that if he aided his fellow-men, he would commonly receive aid in return" (Darwin 1871, 1, 163). Critics, like David Stove (2007), who complain that Darwinism means constant struggle – "If his theory or explanation of evolution were true, there would be in every species a constantly recurring struggle for life: a competition to survive and reproduce which is so severe that few of the competitors in any generation can win" (45) – simply do not know what they are talking about. Of course, the real hope of Stove and his comrades is that you do not know what Darwin is talking about.

This does not mean that we are always nice to everyone all the time. Obviously not! My worries at the end of the last chapter. Hatred, prejudice, violence, crime, war. Those, like me, who have lived so long in such peaceable societies, wonder constantly at their luck. Why are these horrible phenomena so common? In broad outline, the evolutionary answer is obvious. Darwin knew. Individual selection! "It is no argument against savage man being a social animal, that the tribes inhabiting adjacent districts are almost always at war with each other; for the social instincts never extend to all the individuals of the same species" (Darwin 1871, 1, 85). The question really is not so much that tribes (that Darwin took to be groups of interrelated individuals) would be wary of other tribes, but precisely why this would happen. Clearly, it is either because they see the others as a threat, or because the others have things that they want. Agriculture! Moreover, in the evolutionary world, one always suspects that sex is not far from the surface – even if we are not all Don Giovanni – and one presumes that the women of others would be an attraction, as conversely your women would be an attraction to others. One thinks of the abuse of German women as the Russians started to invade Germany, East Prussia particularly. This was abuse sanctioned – encouraged – by the higher echelons of the Russian military.

Against this, notice how we have as part of our nature ways of dealing with these issues. We restrain violence thanks to strategies like just war theory. Obliteration bombing, indiscriminate bombing of civilians – notwithstanding all sides did it in the Second World War – is wrong. Coventry should not have been destroyed. Dresden should not have been destroyed. Hiroshima and Nagasaki should not have been destroyed. Sociality is important, and we have ways to promote and safeguard it, even if those ways are not always successful.

Normative Darwinism

> Moral systems are sets of interlocking values, virtues, norms, practices, identities, institutions, technologies, and evolved psychological mechanisms that work together to suppress or regulate self-interest and make cooperative societies possible. (Haidt 2012, 314)

Agreed. Start now with questions to do with substantive ethics. We are subject to the suzerainty of the machine metaphor, so Hume's is/ought distinction reigns supreme. The fact that we do these things does not in itself make them moral. However, what is important is that so many of the things we do and think are precisely those things that we would expect of moral people, and, when they are not what we would expect of moral people, we have the conceptual tools to realize why they are not what we would expect of moral people. Morally, we realize we should be kind to small children. Morally, we realize that we are in the wrong if we do not pay our share into the coffee fund. John Rawls in his celebrated *A Theory of Justice* makes this point.

> In arguing for the greater stability of the principles of justice I have assumed that certain psychological laws are true, or approximately so. I shall not pursue the question of stability beyond this point. We may note however that one might ask how it is that human

> beings have acquired a nature described by these
> psychological principles. The theory of evolution
> would suggest that it is the outcome of natural
> selection; the capacity for a sense of justice and the
> moral feelings is an adaptation of mankind to its place
> in nature. (Rawls 1971, 502–3)

Rawls, like Plato in the *Republic*, is offering a contract theory –
what's a good setup for a group of folk living together? One of
the problems of such a theory is that, historically, it never
seems very plausible that a group of wise men (and perhaps
women) sat around and drew up the rules for proper conduct.
It makes more sense to leave it to natural selection.

Still, an urge to be nice to small children is not a
moral claim – you *ought* to be nice to small children. Rawls
appreciates this, noting that even if biology does underlie the
contract theory, it doesn't follow that it is morally obligatory
to follow it. "These remarks are not intended as justifying
reasons for the contract view" (Rawls 1971, 504). At most we
seem to have a Darwinized version of the emotive theory of
ethics (Ayer 1936). I don't like being unkind to small children.
Boo-hoo! I don't like it when you are unkind to small children. For my peace of mind, I urge you not to be unkind to
small children. Let us agree that in Darwin's world, moral
sentiments are emotions. Very much akin to the categories/
habits in epistemology, they are put in place to help us get on
in life. However, and this is very Humean, phenomenologi-
cally as it were, moving on beyond raw emotivism, they must
present themselves as more than mere emotions, desires.
If they are just that, then because of cheating they are
going to break down. Why should I respect your marriage?

Why should you likewise respect my marriage? Canadian philosophy departments in the early years of the pill.

We must have some restraint on the naked emotions, and it is here that morality comes in. To use an ugly word of the late J. L. Mackie (1977), we "objectify" our emotions – they present themselves as objectively binding. You ought not violate the marriage bonds of yourself or of others. This doesn't mean you never will, rather that you ought not. And because of the overall force of this moral prohibition, a kind of societal stability prevails.

Clarifications

Putting things together, at the substantive or normative level, Darwinian evolutionary theory generates the norms of morality. The norms of morality that we accept as, let us say, common sense. You ought to care about children, you ought not cheat on your wife sort of thing. Am I, almost deliberately, ignoring what philosophers through the ages have had to say about normative ethics? What about the Categorical Imperative? What about the Greatest Happiness Principle? Do these have no place in my world picture? In the most important way, my position is intended to encompass them all! Philosophers make a living out of finding awkward counterexamples, but all the traditional suggestions agree on most issues. A Kantian, with thoughts about ends in themselves, would be horrified if – Nazi style – you intended to do horrendous operations on small children to make important medical discoveries. A utilitarian, especially of the John Stuart Mill ilk, would be equally horrified. Even if you get important discoveries in this case, the unhappiness of the child is all-important, as is the sense of certainty that children (including your own) are not going to be whisked away in the cause of medical science.

What of those awkward counterexamples? To take one of those ethical dilemmas on which so many cases for tenure are built, what should we do if we find ourselves faced with a runaway railway truck, with six people tied up just downline (Thomson 1985). Should we pull the switch and divert the truck to a sideline on which only one person is tied up? What if there is no switch but we are standing on a bridge next to a fat man – he must be big, so self-sacrifice is not an option – and we can push him over, thereby killing him but stopping the truck? Intuition usually tells us that it is okay to pull the switch but not to upend our overweight neighbor. Why? Formally, they are identical. The Darwinian has little time for any of this. The answer lies not in ourselves but in our genes. In our past, we have often had situations where we must make decisions about the welfare of our neighbors. Selection has programed us to care about them, if only because we are their neighbors and we need them to care about us. In our past, rarely if ever have we had the problem of pulling a switch. We can avoid emotions and make a rational decision. We are going to act differently in different circumstances. Don't look to evolution to offer a unique proper solution. I regard this as a strength of the Darwinian case, not a weakness.

I spoke above of the hostility we have to outsiders, to the other. When violence occurs, particularly in the

context of war, like the complexities of thinking epistemo-
logically, we are primed to think through the complexities
of morality. Our moralists spring into action, showing why
and how violence can at times be justified. Agreed, but one
final question. We are not hostile to outsiders all the time.
I don't hate or fear the folk from Timbuctoo. If there is no
reason to feel hostility to outsiders, does this then mean we
are neutral towards them, or, as Jesus rather implied,
should we extend our moral concerns out to everyone?
Darwin thought this too. For all that he thought violence
towards others rooted in our past, he thought we could
overcome our legacy. Remember how he told us that gradually we reach out from our close group to society at large,
and then "there is only an artificial barrier to prevent his
sympathies extending to the men of all nations and races"
(Darwin 1874, 122). Perhaps so. My sense, however, is that,
once again, Hume was right about these sorts of things.
Morality works on a gradient.

> A man naturally loves his children better than his
> nephews, his nephews better than his cousins, his cousins
> better than strangers, where everything else is equal.
> Hence arise our common measures of duty, in preferring
> the one to the other. Our sense of duty always follows the
> common and natural course of our passions. (Hume
> 1739–40, 483–84)

Although there are suggestions in the Bible, for instance the
story of the Good Samaritan, that we might have equal
obligations to all, more generally it agrees with Hume.
"Anyone who does not provide for their relatives, and especially for their own household, has denied the faith and is

worse than an unbeliever" (I Timothy 5:8). The Darwinian thinks likewise.

Darwinian Metaethics

In a world of the naturalistic fallacy, what then of metaethics? You cannot justify Darwinian normative ethics by the fact of Darwinian evolution. G. E. Moore appealed to what he called non-natural properties. The rules of morality exist in a kind of ethereal world, as do the rules of mathematics. If this sounds Platonic, it was intended so. "I am pleased to believe that this is the most Platonic system of modern times" (Baldwin 1990, 50). Speaking to an issue I brushed past at the beginning of the last chapter, the Darwinian is not very comfortable with this. For a Darwinian ethicist like me, the non-progressive nature of the evolutionary process comes to the fore. In the case of epistemology – knowledge about the world, for better or for worse — our habits come together to say that there is a train bearing down on us. If our habits didn't do this, it would seem (to us using those habits of reasoning we do have) we are not long for this world. In the ethical case, you could have different systems – systems with rival demands – and they could all work. Darwin himself noted this.

> I do not wish to maintain that any strictly social animal, if its intellectual faculties were to become as active and as highly developed as in man, would acquire exactly the same moral sense as ours. In the same manner as various animals have some sense of beauty, though they admire widely different objects, so they might have a sense of right and wrong, though led by it to follow widely

different lines of conduct. If, for instance, to take an extreme case, men were reared under precisely the same conditions as hive-bees, there can hardly be a doubt that our unmarried females would, like the worker-bees, think it a sacred duty to kill their brothers, and mothers would strive to kill their fertile daughters; and no one would think of interfering. Nevertheless the bee, or any other social animal, would in our supposed case gain, as it appears to me, some feeling of right and wrong, or a conscience. (Darwin 1871, 1, 73)

We are humans not hive-bees and, thank God or Darwin, we males do not have to worry as winter approaches. But the relativity point is made. Platonic ideals don't have a dog in this fight.

Could the relativity point be made about humans as humans? Could we have different systems here? Not within the species. We saw in an earlier chapter, there is a lot of uniformity about humans. For all that I sometimes have down days – my wife is a very strong-minded woman – truly we are not hymenoptera. I see absolutely no reason to suppose men and women have or obey different moral codes. But what about the human species taken as a whole? Could it have different systems? Kant (1785) seemed to think so in some sense. At least, he thought it logically possible that we have a society without morality, where everything is done rationally – I will give to you but what will you give to me? He just didn't think it could work. And he has a point. Apart from anything else, time is money. We couldn't function if we had to calculate at every point. The child runs in front of the bus. Can I save it?

What are the chances of my getting hurt? Does it really matter if the kid is killed? Am I not more valuable than the kid? What will the parents give me if I save their kid? By the time I have made the calculations, the kid will be like the wretched Rebecca who slammed doors until she upset a statue above her: "It knocked her flat! It laid her out! She looked like that."

Note, the objection is pragmatic. It is not that using just reason is wrong. It wouldn't work. Could we have an alternative functioning morality? Suppose we had the John Foster Dulles system of morality (Ruse 1986). He was Eisenhower's secretary of state during the Cold War. He hated the Russians. He believed he ought to hate the Russians. But he knew that they felt the same way about him. So, they got along. Judged by today's standards, not too badly either. Notice what this all means. There is no direction to evolution. This means that we could have the Dulles morality, thinking that hating others is morally obligatory. We go all the way – born, live, die – with this. We are in total ignorance of the true objective morality. God's will, Platonic forms or whatever: "But I tell you, love your enemies and pray for those who persecute you" (Matthew 5:44). Somehow this ignorance strikes me as a reductio of objective morality. I can accept that one or two psychopaths might be unable to perceive objective morality, but all of us? Objective morality must in some sense reach out to normal human beings. Plato, of all people, would be horrified if it did not. Hence, the non-directedness of Darwinian evolution means that there is no objective morality. Or, as it is sometimes put, it "debunks" the case for moral realism, directing you to moral non-realism. In that sense, there is no justification – meaning outside justification – for substantive ethics.

One final question. Should I be telling you all of this? Now that I have given the game away, why not go out and rape and pillage to your heart's content? At least, do whatever you want while being careful not to get caught. There is no objective morality, so make hay while the sun shines. The trouble is – or rather the good thing is – humans don't work that way. We have evolved to be part of the system. We cannot deny human nature even if we want to. Hume again. He is talking of the skepticism induced by his study of causation. "I dine, I play a game of backgammon, I converse, and am merry with my friends. And when, after three or four hours' amusement, I would return to these speculations, they appear so cold, and strained, and ridiculous, that I cannot find in my heart to enter into them any farther" (Hume 1739–40, 175). Same with morality. What our reason tells us is going to be capped by our emotions. In *Crime and Punishment*, Dostoevsky tells of the student Raskolnikov who thinks he can escape conventional morality and who commits a murder. The police chief knows that he did it; but, knowing that he will not be able to live with himself, waits until the pressure becomes too great and, of his own free will, the student confesses. Perhaps, for a day or two, after reading this book, you will go out and do what you please. God help the library books because apparently Darwin won't. You'll be sorry! Sleepless nights, wracked by conscience, at that yellow highlighting of the *Critique of Pure Reason*. How could you destroy the joy of others as they set out, all pure and innocent and eager, for the first time into the Transcendental Analytic, to find the Metaphysical Deduction has been marked up like a copy-editor's proof? And that awful question is a good note on which to end this chapter.

Epilogue

What about human beings, in the moral realm, as seen through the lenses and demands of the three groups of the first chapter – the religious, the secular, the creationist?

Religious? I am now going to take it for granted that a mainline Christian (and people of other religions) can and does accept that human beings are the product of evolution by natural causes. There are tensions, especially for the mechanist/Darwinian, particularly over the necessary appearance of human beings. People who take faith as all-important are going to be supremely unbothered. "Now we see through a glass darkly." (KJV!) For those who do think reason must have a larger role, I have made some suggestions for their resolution, although whether – in the light of such things as quantum indeterminacy – they are adequate is, as they say, a problem left to the reader. The organicist is going to feel fewer tensions, especially with respect to the necessary appearance of humans. It is all pre-programed anyway. There is not the randomness of Darwinism. Indeed, all along, the organicist is going to be somewhat smug, explaining why so many of the religious embrace organicism. Evolution leads up to human beings, and part of their high level is that they can appreciate objective right or wrong. This holds whether the right and wrong is directly the will of God – like the God of Job: "What I say goes" – or whether God has created Platonic-like rules for us to read and obey – like the rules of mathematics. The metaphor insists that values emerge from the process – there is a proper direction and end – and that direction and end makes no sense if the ultimate beings cannot discern right and wrong, even if we might not yet be totally perfect and still have trouble making out all of the details.

You might think that, even apart from issues mentioned just above, the mechanist, the Darwinian, is in real trouble here. The Darwinian is a moral non-realist and that is surely the last thing we can be. As an authority close to my heart has said: "morality is an illusion put in place by our genes to make us good cooperators" (Ruse and Wilson 1986). One thing you can say with certainty. For the God of the Bible, Old and New Testaments, morality is not an illusion. Ask David who got roughed up for going after another man's wife, Bathsheba. Ask Judas, who felt so awful he went out and hanged himself. Stop for a moment. The illusion, if such there be, lies at the metaethical level not the substantive level. The classic theological treatment of the justification of ethics – natural law theory – does not say that there is some non-natural code written in the rational world – a kind of spiritual tablet on which the ten commandments were written. It says that God's will, doing what is right, is doing what is natural. Caring for little children is natural and so is good. What Jeffrey Dahmer was doing was non-natural and hence bad. The Darwinian says that we humans have evolved naturally with our moral code. Isn't this what God wanted and is good? There is of course the problem of what if we had – what if we have – evolved in such a way as to have what God thinks is a bad system. Natural for us, bad in

God's eyes. The John Foster Dulles code of morality. I guess this is the problem mentioned already. God in His wisdom looked at the options and actualized the one He wanted. Assuming He could do that.

In the secular world, little more needs to be said. The organicists are going to be happy. Evolution progressed to the top dog, humans, and part of being human is recognizing in some sense what is right and what is wrong. Not the will of God but in some sense objective, found not created. Darwinians of course can be secularists, often/usually are. But the natural world cannot prove human superiority. That involves absolute value judgments and the mechanist qua mechanist cannot make these. Really, the Darwinian is directed towards (small c) creationism. Existentialism. We make our own values. If we are to make claims about the superiority of humans, they come from us and not elsewhere. The Darwinian must go this way. Interestingly, there are those who have claimed to be Christian existentialists. Obviously not of the Sartrean kind, for whom the existence of God is irrelevant. More of the Kierkegaardian kind, seeing no ultimate guidance in reason but all a matter of faith and commitment. Were I a Christian, I would be drawn this way, particularly in the moral world. My whole life has been molded by the Quaker doctrine of the earlier-mentioned "inner light," the belief that humans are special because in some way we all are touched by the divine. As a non-believer, I don't accept it literally, and I have made it clear that I have trouble extending it to all people without exception, or at least qualification. Overall though, as a teacher, without intending to sound prissy, I have been guided and inspired and helped by the starting assumption that each and every one of my students, no matter how difficult or psychically unattractive, is a person of worth to whom I, as a fellow human, have obligations.

Isn't the would-be Darwinian existentialist going to run into trouble? For the creationist/existentialist, the rules themselves are part of the choice, not just what you do when you have the rules. Sartre (1948) tells a story.

> I will quote the case of one of my pupils who came to me. He lived alone with his mother, his father having gone off as a collaborator and his brother killed in 1940. He had a choice – to go and fight with the Free French to avenge his brother and protect his nation, or to stay and be his mother's only consolation. So he was confronted by two modes of action; one concrete and immediate but directed only towards one single individual; the other addressed to an infinitely greater end but very ambiguous. What would help him choose? Christian doctrine? Accepted morals? Kant? (3)

Sartre tells us that the answer has to lie within ourselves. We in a way create the way forward.

> I said to him, "In the end, it is your feelings which count." But how can we put a value on a feeling?
> At least, you may say, he sought the counsel of a professor. But, if you seek advice, from a priest for example, in choosing which priest you know already, more or less, what they would advise. (3)

The existentialist doesn't buy into that overused mantra of the sixties: "What you want is what you like is what is okay."

What you like is sometimes very much not okay. Jeffrey Dahmer. But somehow you must see that the choice lies within you, not relying on external authority for instructions.

> Let us say that moral choice is comparable to a work of art. Do we reproach the artist who makes a painting without starting from laid-down rules? Did we tell him what he must paint? There is no pre-defined picture, and no-one can say what the painting of tomorrow should be; one can judge only one at a time.
>
> Amongst morals, the creative situation is the same, and just as the works of, say, Picasso, have consequences, so do our moral judgements. (5)

We can make judgments about works of art. We can make judgments about moral decisions. Yet, don't we run into the roadblock that the Darwinian believes that we are born with a moral nature and innate moral rules? It was not I who decided that it is wrong to kill people and eat them. It was not I who decided that I was in Mother Teresa mode when I spent an hour with that needy student, rather than joining my pals in the faculty club to gossip about the dean. There is a ready answer to these kinds of issues. Sartre misses the extent to which being a human being is a group experience. No person is an island. Morality comes from within human beings, individually and collectively. We are social, descended from a very small group. That doesn't mean that *Homo sapiens* is a kind of super-organism, with us all as parts (Ruse 1987). We individually work towards our own ends, but as part of a network in which we are interdependent. Our ancestors made choices about right and wrong. It is wrong to hurt little children, even those that are not ours. We inherit this from them, so we don't have to do it all ourselves. But, at the same time, we individually must accept (or reject) our heritage. Those guards at Auschwitz who killed little children were vile, they were wrong. Not because of rules coming from some neo-Platonic heaven of pure rationality, or because of some inherent part of nature, but because we human beings made that choice. I inherited but at the same time I affirmed. "The mutual dependence of men is so great in all societies that scarce any human action is entirely complete in itself, or is performed without some reference to the actions of others, which are requisite to make it answer fully the intention of the agent" (Hume 1748).

Do I really have that much freedom? I – we – face challenges. I claim to be a pragmatist, so obviously I cannot do anything I might want. Tempting though it be at times, I cannot kill off all my children as soon as they reach adolescence. But isn't this David Hume's point? Isn't this Jonathan Edwards's point? Having no restraints is not freedom. It is craziness. Freedom is a matter of negotiating a successful way forward and doing it myself. I – my group – have chosen cooperation through kindness rather than, Dulles style, cooperation through fear and hatred. Is it better? It is mine and I think it better. Ultimately, though, there is no absolute standard, no absolute values. Pragmatism! It may be that my way works better than the Dulles way. All I can say is that I and my group have chosen our way. I suspect that, even if the Dulles system theoretically works as well as or better (from a biological standpoint) than our system, and although admittedly there can be great satisfaction in diddling an opponent, overall we are a lot happier with our system. John Stuart Mill knew whereof he wrote. Being happy pays off. For those who have the guts to go out and grab the opportunity, our human nature leads to a life of huge personal satisfaction. There is no higher or lower standard. No ultimate court of appeal. This not a message of despair or failure. It is one of great hope and excitement.

There is my answer to the question: Are humans special? More accurately: Why do we think we are special? The religious get their answer from their faith. The secular from subscribing to the organic metaphor. The Darwinian, a mechanist, looks within. It is having to stand on your own – the opportunities, the challenges, the choices – that makes humans of great worth. As a Darwinian existentialist, I ask for no more than I have. As a Darwinian existentialist, I want no more than I have. I am so privileged to have had the gift of life and the abilities and the possibilities to make full use of it.

BIBLIOGRAPHY

Agricola, G. [1556] 1950. *De Re Metallica*. Translators H. C. Hoover and L. C. Hoover. New York, NY: Dover.

Allison, A. C. 1954a. Protection afforded by sickle-cell trait against subtertian malarial infection. *British Medical Journal* 1: 290–94.

Allison, A. C. 1954b. The distribution of the sickle-cell trait in East Africa and elsewhere, and its apparent relationship to the incidence of subtertian malaria. *Transactions of the Royal Society of Tropical Medical Hygiene* 48: 312–18.

Aquinas, St. T. [1259–65] 1975. *Summa contra Gentiles*. Translator V. J. Bourke. Notre Dame, IN: University of Notre Dame Press.

Aquinas, St. T. [1265–74] 1981. *Summa Theologica*. Translators Fathers of the English Dominican Province. London: Christian Classics.

Augustine, St. 1991. *On Genesis*. Translator R. J. Teske. Washington, D.C.: Catholic University of America Press.

Ayer, A. J. 1936. *Language, Truth and Logic*. London: Gollancz.

Bacon, F. [1605] 1868. *The Advancement of Learning*. Oxford: Clarendon Press.

Bada, J. L., and A. Lazcana. 2009. The origin of life. In *Evolution: The First Four Billion Years*. Editors M. Ruse and J. Travis, 49–79. Cambridge, MA: Harvard University Press.

Bakker, R. T. 1983. The deer flees, the wolf pursues: Incongruencies in predator–prey coevolution. In *Coevolution*. Editors D. J. Futuyma and M. Slatkin, 350–82. Sunderland, MA: Sinauer.

Baldwin, T. 1990. *G. E. Moore*. London: Routledge and Kegan Paul.

Barnes, J., editor. 1984. *The Complete Works of Aristotle*. Princeton, NJ: Princeton University Press.

Bennett, M. J., and T. S. Posteraro, editors. 2019. *Deleuze and Evolutionary Theory*. Edinburgh: University of Edinburgh Press.

Benton, M. J. 2009. Paleontology. In *Harvard Companion to Evolution*. Editors M. Ruse and J. Travis, 80–104. Cambridge, MA: Harvard University Press.

Bergson, H. 1907. *L'évolution créatrice*. Paris: Alcan.

Bergson, H. 1911. *Creative Evolution*. New York, NY: Holt.

Bowler, P. 1989. *The Mendelian Revolution: The Emergence of Hereditarian Concepts in Modern Science and Society*. London: The Athlone Press.

Boyle, R. [1686] 1996. *A Free Enquiry into the Vulgarly Received Notion of Nature*. Editors E. B. Davis and M. Hunter. Cambridge: Cambridge University Press.

Boyle, R. [1688] 1966. A disquisition about the final causes of natural things. In *The Works of Robert Boyle*. Editor T. Birch, Vol. 5: 392–444. Hildesheim: Georg Olms.

Broad, C. D. 1944. Critical notice of Julian Huxley's *Evolutionary Ethics*. *Mind* 53: 344–67.

Browne, J. 1995. *Charles Darwin: Voyaging. Volume 1 of a Biography*. London: Jonathan Cape.

Browne, J. 2002. *Charles Darwin: The Power of Place. Volume 2 of a Biography*. London: Jonathan Cape.

Bury, J. B. [1920] 1924. *The Idea of Progress: An Inquiry into Its Origin and Growth*. London: Macmillan.

Carnegie, A. 1889. The gospel of wealth. *North American Review* 148: 653–65.

Clifford, W. K. 1901. Body and mind. In *Lectures and Essays of the Late William Kingdom Clifford*. Editors L. Stephen and F. Pollock, Vol. 2: 1–51. London: Macmillan.

Birch, C., and J. B. Cobb Jr. 1981. *The Liberation of Life*. Cambridge: Cambridge University Press.

Conway Morris, S. 2003. *Life's Solution: Inevitable Humans in a Lonely Universe*. Cambridge: Cambridge University Press.

Cooper, J. M., editor. 1997. *Plato: Complete Works*. Indianapolis, IN: Hackett.

Coyne, J. A. 2012. Another philosopher proclaims a nonexistent "crisis" in evolutionary biology. *Why Evolution Is True*, September 12. https://whyevolutionistrue.com/2012-09/07/another-philosopher-proclaims-a-nonexistent-crisis-in-evolutionary-biology/.

Coyne, J. A. 2015. *Faith versus Fact: Why Science and Religion Are Incompatible*. New York, NY: Viking.

Coyne, J. A., N. H. Barton, and M. Turelli. 1997. Perspective: A critique of Sewall Wright's shifting balance theory of evolution. *Evolution* 51: 643–71.

Darwin, C. 1851. *A Monograph of the Sub-class Cirripedia, with Figures of All the Species. The Lepadidae; or Pedunculated Cirripedes*. London: Ray Society.

Darwin, C. 1859. *On the Origin of Species by Means of Natural Selection, or the Preservation of Favoured Races in the Struggle for Life*. London: John Murray.

Darwin, C. 1861. *Origin of Species*. Third edition. London: John Murray.

Darwin, C. 1871. *The Descent of Man, and Selection in Relation to Sex*. London: John Murray.

Darwin, C. 1874. *The Descent of Man (Second Edition)*. London: John Murray.

Darwin, C. 1958. *The Autobiography of Charles Darwin, 1809–1882*. Editor N. Barlow. London: Collins.

Darwin, C. 1985–. *The Correspondence of Charles Darwin*. Cambridge: Cambridge University Press.

Darwin, C. 1987. *Charles Darwin's Notebooks, 1836–1844*. Editors P. Barrett, P. Gautrey, S. Herbert, D. Kohn, and S. Smith. Ithaca, NY: Cornell University Press.

Darwin, E. [1794–96] 1801. *Zoonomia; or, The Laws of Organic Life*. Third edition. London: J. Johnson.

Darwin, E. 1803. *The Temple of Nature*. London: J. Johnson.

Dawkins, R. 1976. *The Selfish Gene*. Oxford: Oxford University Press.

Dawkins, R. 1986. *The Blind Watchmaker*. New York, NY: Norton.

Dawkins, R. 2003. *A Devil's Chaplain: Reflections on Hope, Lies, Science and Love*. Boston, MA and New York, NY: Houghton Mifflin.

Dawkins, R. 2006. *The God Delusion*. New York, NY: Houghton, Mifflin, Harcourt.

Dawkins, R., and J. R. Krebs. 1979. Arms races between and within species. *Proceedings of the Royal Society of London, Series B: Biological Sciences* 205: 489–511.

Descartes, R. [1637] 1964. Discourse on method. In *Philosophical Essays*, 1–57. Indianapolis, IN: Bobbs-Merrill.

Descartes, R. [1644] 1955. The principles of philosophy. In *The Philosophical Works of Descartes*. Translators E. Haldane and G. R. T. Ross, Vol. 1: 201–302. New York, NY: Dover.

Diderot, D. 1943. *Diderot: Interpreter of Nature*. Editors J. Kemp and J. Stewart. New York, NY: International Publishers.

Dietl, G. P. 2003. Coevolution of a marine gastropod predator and its dangerous prey. *Biological Journal of the Linnean Society* 80: 409–36.

Dijksterhuis, E. J. 1961. *The Mechanization of the World Picture*. Oxford: Oxford University Press.

Dobzhansky, T. 1937. *Genetics and the Origin of Species*. New York, NY: Columbia University Press.

Duncan, D., editor. 1908. *Life and Letters of Herbert Spencer*. London: Williams and Norgate.

Dupré, J. 2003. *Darwin's Legacy: What Evolution Means Today*. Oxford: Oxford University Press.

Dupré, J. 2010. The conditions for existence. *American Scientist* 98: 170.

Dupré, J. 2012a. *Processes of Life: Essays in the Philosophy of Biology*. Oxford: Oxford University Press.

Dupré, J. 2012b. Evolutionary theory's welcome crisis. *Project Syndicate*. www.project-syndicate.org/commentary/evolutionary-theory-s-welcome-crisis-by-john-dupre?barrier=accesspaylog.

Dupré, J. 2017. The metaphysics of evolution. *Interface Focus*. https://doi.org/10.1098/rsfs.2016.0148.

Edwards, A. W. F. 2003. Human genetic diversity: Lewontin's fallacy. *BioEssays* 25: 798–801.

Edwards, J. [1754] 2013. *A Careful and Strict Enquiry into the Modern Prevailing Notions of That Freedom of Will, Which Is Supposed to Be Essential to Moral Agency, Vertue and Vice, Reward and Punishment, Praise and Blame*. Overland Park, KS: Digireads.

Fichte, J. G. [1821] 1922. *Addresses to the German Nation*. Chicago, IL: Open Court.

Fisher, R. A. 1930. *The Genetical Theory of Natural Selection*. Oxford: Oxford University Press.

Fodor, J. 2007. Why pigs don't have wings. The case against natural selection. *London Review of Books* 29 (20) 18 October.

Fodor, J., and M. Piattelli-Palmarini. 2010. *What Darwin Got Wrong*. New York, NY: Farrar, Straus, and Giroux.

Ford, E. B. 1964. *Ecological Genetics*. London: Methuen.

Frost, R. 1931. Education by poetry: A meditative monologue. *Amherst Graduates' Quarterly* XX: 75–85.

Futuyma, D. J. 2017. Evolutionary biology today and the call for an extended synthesis. *Interface Focus, Royal Society Publishing*, 7. https://doi.org/10.1098/rsfs.2016.0145.

Gare, A. 2002. The roots of postmodernism: Schelling, process philosophy and poststructuralism. In *Process and Difference: Between Cosmological and Poststructuralist Postmodernisms*. Editors C. Keller and A. Daniell, 31–54. Albany, NY: SUNY Press.

Gibson, A. 2013. Edward O. Wilson and the organicist tradition. *Journal of the History of Biology* 46: 599–630.

Gilbert, S. F., J. M. Opitz, and R. A. Raff. 1996. Resynthesizing evolutionary and developmental biology. *Developmental Biology* 173: 357–72.

Gould, S. J. 1977. *Ontogeny and Phylogeny*. Cambridge, MA: Belknap Press.

Gould, S. J. 1981. *The Mismeasure of Man*. New York, NY: Norton.

Gould, S. J. 1985. SETI and the wisdom of Casey Stengel. *The Flamingo's Smile*, 403–13. New York, NY: Norton.

Gould, S. J. 1988. On replacing the idea of progress with an operational notion of directionality. In *Evolutionary Progress*. Editor M. H. Nitecki, 319–38. Chicago, IL: University of Chicago Press.

Gould, S. J., and N. Eldredge. 1977. Punctuated equilibria: The tempo and mode of evolution reconsidered. *Paleobiology* 3: 115–51.

Gould, S. J., and R. C. Lewontin. 1979. The spandrels of San Marco and the Panglossian paradigm: A critique of the adaptationist programme. *Proceedings of the Royal Society of London, Series B: Biological Sciences* 205: 581–98.

Greene, J. 2013. *Moral Tribes: Emotion, Reason, and the Gap between Us and Them*. New York, NY: Penguin.

Greene, J. C., and M. Ruse. 1996. On the nature of the evolutionary process: The correspondence between Theodosius Dobzhansky and John C. Greene. *Biology and Philosophy* 11: 445–91.

Haas, J., and M. Piscitelli. 2013. The prehistory of warfare: Misled by ethnography. In *War, Peace, and Human Nature: The Convergence of Evolutionary and Cultural Views*. Editor D. P. Fry, 168–90. Oxford: Oxford University Press.

Haeckel, E. 1896. *The Evolution of Man*. New York, NY: Appleton.

Haidt, J. 2012. *The Righteous Mind: Why Good People Are Divided by Politics and Religion*. New York, NY: Vintage.

Haldane, J. B. S. 1927. *Possible Worlds and Other Essays*. London: Chatto and Windus.

Haldane, J. B. S. 1932. *The Causes of Evolution*. New York, NY: Cornell University Press.

Hamilton, W. D. 1964a. The genetical evolution of social behaviour I. *Journal of Theoretical Biology* 7: 1–16.

Hamilton, W. D. 1964b. The genetical evolution of social behaviour II. *Journal of Theoretical Biology* 7: 17–32.

Harari, Y. N. 2015. *Sapiens: A Brief History of Humankind*. New York, NY: Harper.

Harvey, P. 1990. *An Introduction to Buddhism: Teachings, History and Practices*. Cambridge: Cambridge University Press.

Hegel, G. W. F. [1817] 1970. *Philosophy of Nature*. Oxford: Oxford University Press.

Hopson, J. A. 1977. Relative brain size and behavior in archosaurian reptiles. *Annual Review of Ecology and Systematics* 8: 429–48.

Hume, D. [1739–40] 1978. *A Treatise of Human Nature*. Oxford: Oxford University Press.

Hume, D. [1748] 2007. *An Enquiry Concerning Human Understanding*. Oxford: Oxford University Press.

Huxley, J. S. 1912. *The Individual in the Animal Kingdom*. Cambridge: Cambridge University Press.

Huxley, J. S. 1927. *Religion without Revelation*. London: Ernest Benn.

Huxley, J. S. 1934. *If I Were Dictator*. New York, NY and London: Harper and Brothers.

Huxley, J. S. 1942. *Evolution: The Modern Synthesis*. London: Allen and Unwin.

Huxley, T. H. 1863. *Evidence as to Man's Place in Nature*. London: Williams and Norgate.

Huxley, T. H. [1893] 2009. *Evolution and Ethics with a New Introduction*. Editor M. Ruse. Princeton, NJ: Princeton University Press.

Jerison, H. 1973. *Evolution of the Brain and Intelligence*. New York, NY: Academic Press.

Johanson, D., and M. Edey. 1981. *Lucy: The Beginnings of Humankind*. New York, NY: Simon and Schuster.

Kant, I. [1787] 2017. *Critique of Pure Reason*. Second edition. Translator and editor J. Bennett. www.earlymoderntexts.com/assets/pdfs/kant1781part1.pdf.

Kant, I. 1783. *Prolegomena*. http://web.mnstate.edu/gracyk/courses/phil%20306/kant_materials/prolegomena1.htm#info.

Kant, I. [1785] 1959. *Foundations of the Metaphysics of Morals*. Indianapolis, IN: Bobbs-Merrill.

Kant, I. [1790] 2000. *Critique of the Power of Judgment*. Editor P. Guyer. Cambridge: Cambridge University Press.

Kepler, J. [1619] 1977. *The Harmony of the World*. Translators E. J. Aiton, A. M. Duncan, and J. V. Field. Philadelphia, PA: American Philosophical Society.

Kierkegaard, S. 1992. *Concluding Unscientific Postscript to Philosophical Fragments, Volume 1 (Kierkegaard's Writings, Vol. 12.1)*. Translators H. V. Hong and E. H. Hong. Princeton, NJ: Princeton University Press.

Kuhn, T. 1957. *The Copernican Revolution*. Cambridge, MA: Harvard University Press.

Kuhn, T. 1962. *The Structure of Scientific Revolutions*. Chicago, IL: University of Chicago Press.

Kuhn, T. 1993. Metaphor in science. In *Metaphor and Thought*. Second edition. Editor A. Ortony, 533–42. Cambridge: Cambridge University Press.

Lakoff, G., and Johnson M. 1980. *Metaphors We Live By*. Chicago, IL: University of Chicago Press.

Lamarck, J. B. 1815. *Histoire naturelle des animaux sans vertèbres*. Paris: Verdière.

Leibniz, G. F. W. [1714] 1965. *Monadology and Other Philosophical Essays*. New York, NY: Bobbs-Merrill.

Lewontin, R. C. 1972. The apportionment of human diversity. *Evolutionary Biology* 6: 381–98.

Lewontin, R. C. 1974. *The Genetic Basis of Evolutionary Change*. New York, NY: Columbia University Press.

Lewontin, R. C. 2002. *The Triple Helix: Gene, Organism, and Environment*. Cambridge, MA: Harvard University Press.

Lieberman, D. E. 2013. *The Story of the Human Body: Evolution, Health, and Disease*. New York, NY: Vintage.

Lurie, E. 1960. *Louis Agassiz: A Life in Science*. Chicago, IL: Chicago University Press.

Mackie, J. L. 1977. *Ethics*. Harmondsworth: Penguin.

Malthus, T. R. [1826] 1914. *An Essay on the Principle of Population (Sixth Edition)*. London: Everyman.

Maynard Smith, J. 1964. Group selection and kin selection. *Nature* 201: 1145–47.

Maynard Smith, J. 1995. Genes, memes, and minds. *New York Review of Books* 42 (19): 46–48.

Mayr, E. 1942. *Systematics and the Origin of Species*. New York, NY: Columbia University Press.

Mayr, E. 1969. Commentary. *Journal of the History of Biology* 2: 123–28.

McShea, D. W. 1991. Complexity and evolution: What everybody knows. *Biology and Philosophy* 6: 303–25.

McShea, D. W. 1996. Metazoan complexity and evolution: Is there a trend? *Evolution* 50: 477–92.

McShea, D. W., and R. Brandon. 2010. *Biology's First Law: The Tendency for Diversity and Complexity to Increase in Evolutionary Systems*. Chicago, IL: University of Chicago Press.

Mill, J. S. 1863. *Utilitarianism*. London: Parker, Son, and Bourn.

Moore, A. 1890. The Christian doctrine of God. In *Lux Mundi*. Editor C. Gore, 57–109. London: John Murray.

Moore, G. E. 1903. *Principia Ethica*. Cambridge: Cambridge University Press.

Morison, I. 2014. *A Journey through the Universe*. Cambridge: Cambridge University Press.

Murphey, M. G. 1968. Kant's children the Cambridge Pragmatists. *Transactions of the Charles S. Peirce Society* 4: 3–33.

Nagel, T. 2012. *Mind and Cosmos: Why the Materialist Neo-Darwinian Conception of Nature Is Almost Certainly False.* New York, NY: Oxford University Press.

O'Connell, J., and M. Ruse. 2021. *Social Darwinism.* Cambridge: Cambridge University Press.

Owen, R. 1849. *On the Nature of Limbs.* London: Voorst.

Paley, W. [1802] 1819. *Natural Theology (Collected Works: IV).* London: Rivington.

Peirce, C. S. 1877. The fixation of belief. *Popular Science Monthly* 12 (1–15).

Peirce, C. S. 1878. How to make our ideas clear. *Popular Science Monthly* 12: 286–302.

Peirce, C. S. 1958. *Collected Papers of Charles Sanders Peirce.* Cambridge, MA: Harvard University Press.

Pepper, S. C. 1942. *World Hypotheses: A Study in Evidence.* Berkeley, CA: University of California Press.

Peterson, E. L. 2016. *The Life Organic: The Theoretical Biology Club and the Roots of Epigenetics.* Pittsburgh, PA: University of Pittsburgh Press.

Provine, W. B. 1971. *The Origins of Theoretical Population Genetics.* Chicago, IL: University of Chicago Press.

Rawls, J. 1971. *A Theory of Justice.* Cambridge, MA: Harvard University Press.

Reich, D. 2018. *Who We Are and How We Got Here: Ancient DNA and the New Science of the Human Race.* New York, NY: Pantheon.

Richards, R. J. 2003. *The Romantic Conception of Life: Science and Philosophy in the Age of Goethe.* Chicago, IL: University of Chicago Press.

Richards, R. J. 2013. *Was Hitler a Darwinian? Disputed Questions in the History of Evolutionary Theory.* Chicago, IL: University of Chicago Press.

Richards, R. J., and M. Ruse. 2016. *Debating Darwin*. Chicago, IL: University of Chicago Press.

Rorty, R. 1998. *Truth and Progress: Philosophical Papers, III*. Cambridge: Cambridge University Press.

Rosenberg, N. A., J. K. Pritchard, J. L. Weber, H. M. Cann, K. K. Kidd, L. A. Zhivotovsky, and M. Feldman. 2002. Genetic structure of human populations. *Science* 298: 2381–85.

Rousset, B. 1997. La pensée scientifique de Lamarck et la "philosophie de la nature." In *Jean-Baptiste Lamarck, 1744–1829*. Editor G. Laurent, 393–98. Paris: Éditions du CTHS.

Rupke, N. A. 2009. *Richard Owen: Biology without Darwin*. Chicago, IL: University of Chicago Press.

Ruse, M. 1979. *The Darwinian Revolution: Science Red in Tooth and Claw*. Chicago, IL: University of Chicago Press.

Ruse, M. 1986. *Taking Darwin Seriously: A Naturalistic Approach to Philosophy*. Oxford: Blackwell

Ruse, M. 1987. Biological species: Natural kinds, individuals, or what? *The British Journal for the Philosophy of Science* 38: 225–42.

Ruse, M. 1996. *Monad to Man: The Concept of Progress in Evolutionary Biology*. Cambridge, MA: Harvard University Press.

Ruse, M. 2006. *Darwinism and Its Discontents*. Cambridge: Cambridge University Press.

Ruse, M. 2017. *On Purpose*. Princeton, NJ: Princeton University Press.

Ruse, M. 2019. *A Meaning to Life*. Oxford: Oxford University Press.

Ruse, M. 2021. The scientific revolution. In *The Cambridge History of Atheism*. Editors S. Bullivant and M. Ruse. Cambridge: Cambridge University Press.

Ruse, M., and E. O. Wilson. 1985. The evolution of morality. *New Scientist* 1478: 108–28.

Ruse, M., and E. O. Wilson. 1986. Moral philosophy as applied science. *Philosophy* 61: 173–92.

Sartre, J. P. 1948. *Existentialism and Humanism*. Brooklyn, NY: Haskell House Publishers Ltd.

Sidgwick, H. 1876. The theory of evolution in its application to practice. *Mind* 1: 52–67.

Simpson, G. G. 1944. *Tempo and Mode in Evolution*. New York, NY: Columbia University Press.

Simpson, G. G. 1964. *This View of Life*. New York, NY: Harcourt, Brace, and World.

Smith, A. [1776] 1937. *The Wealth of Nations*. New York, NY: Modern Library.

Smith H. M., D. Chiszar, and R. R. Montanucci. 1997. Subspecies and classification. *Herpetological Review* 28: 13–16.

Spencer, H. 1851. *Social Statics: Or, the Conditions Essential to Human Happiness Specified, and the First of Them Developed*. London: Chapman.

Spencer, H. 1852a. A theory of population, deduced from the general law of animal fertility. *Westminster Review* 1: 468–501.

Spencer, H. 1860. The social organism. *Westminster Review* LXXIII: 90–121.

Spencer, H. 1864. *Principles of Biology*. London: Williams and Norgate.

Spencer, H. [1852b] 1868. The development hypothesis. In *Essays: Scientific, Political and Speculative*, 377–83. London: Williams and Norgate.

Spencer, H. [1857] 1868. Progress: Its law and cause. *Westminster Review* LXVII: 244–67.

Spencer, H. 1879. *The Data of Ethics*. London: Williams and Norgate.

Stebbins, G. L. 1950. *Variation and Evolution in Plants*. New York, NY: Columbia University Press.

Stove, D. 2007. *Darwinian Fairytales: Selfish Genes, Errors of Heredity and Other Fables of Evolution*. New York, NY: Encounter Books.

Teilhard de Chardin, P. 1955. *Le phénomène humain*. Paris: Éditions de Seuil.

Templeton, A. R. 2013. Biological races in humans. *Studies in the History and Philosophy of Biology and the Biomedical Sciences* 44: 262–71.

Thomson, J. J. 1985. The trolley problem. *Yale Law Journal* 94: 1395–1415.

Tindal, M. 1732. *Christianity as Old as the Creation: Or, the Gospel, a Republication of the Religion of Nature*. London: n.p.

Von Bernhardi, F. 1912. *Germany and the Next War*. London: Edward Arnold.

Waddington, C. H. 1957. *The Strategy of the Genes*. London: Allen and Unwin.

Waddington, C. H. 1960. *The Ethical Animal*. London: Allen and Unwin.

Waddington, C. H. 1962. *The Nature of Life*. New York, NY: Atheneum.

Watson, J. D., and F. H. C. Crick. 1953. Molecular structure of nucleic acids. *Nature* 171: 737–38.

Whewell, W. 1840. *The Philosophy of the Inductive Sciences*. London: Parker.

Whitcomb, J. C., and H. M. Morris. 1961. *The Genesis Flood: The Biblical Record and Its Scientific Implications*. Philadelphia, PA: Presbyterian and Reformed Publishing Company.

Whitehead, A. N. 1926. *Science and the Modern World*. Cambridge: Cambridge University Press.

Wilson, E. O. 1975. *Sociobiology: The New Synthesis*. Cambridge, MA: Harvard University Press.

Wilson, E. O. 1984. *Biophilia*. Cambridge, MA: Harvard University Press.

Wilson, E. O. 1990. *Success and Dominance in Ecosystems: The Case of the Social Insects*. Oldendorf (Luhe): Ecology Institute.

Wilson, E. O. 1992. *The Diversity of Life*. Cambridge, MA: Harvard University Press.

Wilson, E. O. 2012. *The Social Conquest of Earth*. New York, NY: Norton.

Wright, S. 1931. Evolution in Mendelian populations. *Genetics* 16: 97–159.

Wright, S. 1932. The roles of mutation, inbreeding, crossbreeding and selection in evolution. In *Proceedings of the Sixth International Congress of Genetics* 1: 356–66.

INDEX